ELEMENTS OF FUNCTIONAL
ANALYSIS

ELEMENTS OF FUNCTIONAL ANALYSIS

I. J. MADDOX
Reader in Mathematics
University of Lancaster

CAMBRIDGE
AT THE UNIVERSITY PRESS
1970

Published by the Syndics of the Cambridge University Press
Bentley House, 200 Euston Road, London N.W.1
American Branch: 32 East 57th Street, New York, N.Y.10022

© Cambridge University Press 1970

Library of Congress Catalogue Card Number: 71-85726

Standard Book Number: 521 07617 X

Printed in Great Britain
at the University Printing House, Cambridge
(Brooke Crutchley, University Printer)

This book is dedicated to the memory of
EDITH MADDOX (1877–1958)
May it serve as an epitaph

CONTENTS

PREFACE

There are several excellent books which deal with the subject of functional analysis. Few can be regarded as really elementary or introductory. As a beautiful theory in its own right and for its richness in applications, functional analysis in some shape or form is now taught to second and third year mathematics undergraduates at several British universities. My experience in teaching such students has indicated that they need quite a gentle introduction—largely due to two things: that their analytical abilities are not sufficiently developed and that they are unused to 'abstract' reasoning. In my view, the field of elementary functional analysis is the ideal place in which to learn some abstract structural mathematics and to develop analytical technique.

It is my hope that this book may provide a really introductory, though non-trivial, course on functional analysis for undergraduates who have completed basic courses on real and complex variable theory. Although primarily addressed to students of mathematics it is expected that the approach is basic enough to enable students of physics and engineering to get something of the flavour of the subject.

Of the several excellent books mentioned above, the master work of Banach: *Théorie des opérations Linéaires* (1932) must stand first. Every serious student of analysis should regard his education incomplete until he has read something of this remarkable germinal book.

There is one feature of the present work which we should perhaps mention. Much of the theory is illustrated by examples involving sequence spaces rather than integration spaces. This is partly because most results for sequence spaces will fairly readily generalize to integration spaces, but mainly because the student to whom this book is addressed is unlikely to be sufficiently familiar with integrals of the depth of Lebesgue to enable him to really appreciate examples involving them. However, it has been thought advisable to prove the completeness of the important L_p spaces, referring to works on integration for the relevant theorems on interchange of limit and integral.

Chapter 1 of the book is absolutely fundamental, though extremely elementary. Some may wish to omit it and proceed to the next chapter on metric and topological spaces. In my view it would be best to make certain of the material in chapter 1 before attempting the rest of the text. There are over 300 exercises in the book, many of which are quite routine, though just a few, which appear at the end of the last chapter, are quite difficult. It is recommended that most of the exercises should be attempted—to learn mathematics one must do it.

The final chapter of the book concerns an area of Mathematics which is of special interest to me. Those students who wish to begin graduate work in this field may find it a useful introduction. Readers who are not so inclined may, nevertheless, see functional analysis at work in a fairly concrete situation.

Debts of gratitude are several. At the undergraduate level my interest in analysis was stimulated by Professor D. C. Russell. As a research student I was greatly influenced by my supervisor, Dr B. Kuttner. A number of my colleagues at the University of Lancaster have made helpful comments on the book, during many conversations. I am especially indebted to P. L. Walker for his careful scrutiny of the typescript and for numerous valuable suggestions. Useful assistance was also rendered by J. W. Roles and C. G. Lascarides.

The manuscript was expertly typed by Mrs Sylvia Brennan and Miss June Unsworth, and I gratefully acknowledge their help.

I. J. MADDOX

University of Lancaster, 1969

1

BASIC SET THEORY AND ANALYSIS

1. Sets and functions

The great German mathematician G. Cantor (1845–1918) is usually regarded as the creator of the theory of sets. As our starting point for this book we shall take Cantor's definition of a set: 'A set is any collection of definite, distinguishable objects of our thought, to be conceived as a whole.' The objects mentioned in the definition are called the elements or members of the set. Usually we denote sets by capital letters and elements by lower case letters. If X is a set then we write $x \in X$ to mean that x is an element of X. When an object x is not an element of a set X, we write $x \notin X$.

In what follows we shall take for granted the following sets, which occur throughout mathematics:

$N = \{1, 2, 3, \ldots\}$, the set of all positive integers,

$Z = \{0, 1, -1, \ldots\}$, the set of all integers,

Q, the set of all rational numbers,

R, the set of all real numbers,

C, the set of all complex numbers.

The notation arises as follows: N for natural numbers, Z for Zahlen (German for integers), Q for quotient. The notation R and C seems to require no explanation.

Usually, in a given discussion, we take a fixed set and everything is carried out with reference to it alone. In such a case the fixed set is called the universe of discourse. For example, in number theory the universe of discourse is Z. Within a universe of discourse X a common way of generating a set is to take an object in X of a certain type and then to consider the set of all such objects. For example, having defined an object in Z called a prime number we may then consider the set of all prime numbers.

In a work of the present nature we are primarily concerned with the manipulation of sets, rather than with their deeper properties. To this end we now introduce notation and definitions, and observe some simple results.

[1]

First, there is no way, in general, of explicitly writing down all the elements of a set. For example, it is in the nature of the positive integers N that they cannot all be explicitly exhibited. We have to be content to write $N = \{1, 2, 3, \ldots\}$; the three dots leaving much to the imagination. Generally, we use the curly bracket notation for sets either writing down the first few elements and then some dots which we agree is to tell us that the law of formation of the elements is well-known or obvious, or we put in the law of formation. For example, $\{x \mid x \in N \text{ and } x > 8\}$ is read as 'the set of all x such that x is a positive integer and x is greater than 8'. The vertical bar following x is read as 'such that'. Thus we could write this last set as $\{9, 10, 11, \ldots\}$. Again, $\{x \mid x \in R \text{ and } x > 0\}$ denotes the set of all strictly positive real numbers. In this case it is not possible to write down the elements explicitly, or even in such a way as to indicate the law of formation, such is the nature of the real numbers. In fact it will be seen later that the set $\{x \mid x \in R \text{ and } x > 0\}$ is uncountable, so that the elements cannot even be exhibited as an infinite sequence $x_1, x_2, x_3, \ldots$. We remark that the order of the elements in a set is generally irrelevant. For example, N is the same set as $\{2, 1, 4, 3, 6, 5, \ldots\}$.

If A, B are sets then the notation $A \subset B$ means that every element of A is also an element of B. If $A \subset B$ then we say that A is a subset of B, B is a superset of A, A is included in B and also B includes A. The notation $B \supset A$ is regarded as equivalent to $A \subset B$. We define $A = B$ if and only if $A \subset B$ and $B \subset A$. Also, we say A is a proper subset of B if and only if $A \subset B$ but $A \neq B$. For example, the set of odd integers is a proper subset of Z. We remark that some writers use the notation $A \subseteq B$, which allows equality, and reserve $A \subset B$ for proper subsets. On occasion we shall also say that '$A \subset B$, strictly', meaning that A is a proper subset of B.

Two simple properties of the set inclusion $\subset$ are:

(i) $A \subset A$,

(ii) $A \subset B$ and $B \subset C$ imply $A \subset C$.

If A is a given set let us consider that subset of A defined as $\{x \in A \mid x \neq x\}$. This set has no elements and is known as the *empty set*. It is denoted by $\varnothing$ and has the property that $\varnothing \subset A$ for every set A. Each set $A \neq \varnothing$ has at least two distinct subsets, A and $\varnothing$. If A has only these two subsets then A must be a one element set, $A = \{a\}$, say, where a is the sole element of A. Note that $\varnothing$ has no elements but that the one element set $\{\varnothing\}$ is not empty.

Unions and intersections of sets

Given sets A, B we may form two new sets from them:

$$A \cup B = \{x \mid x \text{ belongs to at least one of } A \text{ and } B\},$$

$$A \cap B = \{x \mid x \in A \text{ and } x \in B\}.$$

We call $A \cup B$ the *union* and $A \cap B$ the *intersection*, of A and B. For example, $\{1, 2, 3\} \cup \{1, 4, 3\} = \{1, 2, 3, 4\}$; $\{2, 3\} \cap \{1, 3, 2\} = \{2, 3\}$. It is trivial that $A \cap B \subset A \subset A \cup B$ for any sets A and B. If $A \cap B = \varnothing$, then we say that A and B are disjoint.

We shall often want to form the union or intersection of a whole class (or collection) of sets. Let $\mathscr{S}$ be a class of sets A. Then we define

$$\cup \{A \mid A \in \mathscr{S}\} = \{x \mid x \in A \text{ for at least one } A \in \mathscr{S}\},$$

$$\cap \{A \mid A \in \mathscr{S}\} = \{x \mid x \in A \text{ for all } A \in \mathscr{S}\}.$$

Sometimes we write $\cup A_\alpha$, $\cap A_\alpha$, where we think of α as running through some indexing set. If α runs through N we usually write

$$\cup \{A_n \mid n \in N\} = \bigcup_{n=1}^{\infty} A_n,$$

and similarly for $\bigcap_{n=1}^{\infty} A_n$. The '$\infty$' in this notation is conventional, but superfluous, not to say confusing. It is emphasized that A_∞ is not in the collection $\{A_n \mid n \in N\}$. Observe also that no limiting process is involved in the above. Thus, for example, to say that $x \in \bigcup_{n=1}^{\infty} A_n$, is to say that there is a positive integer p such that $x \in A_p$.

Example 1. Let A_n be the interval $[0, 1 + 1/n)$ on the real line, i.e. $A_n = \{x \in R \mid 0 \leqslant x < 1 + 1/n\}$, $n = 1, 2, \dots$. Then

$$\bigcap_{n=1}^{\infty} A_n = [0, 1] = \{x \in R \mid 0 \leqslant x \leqslant 1\}.$$

To show this we first prove $[0, 1] \subset \cap A_n$ and then prove $\cap A_n \subset [0, 1]$. Now $x \in [0, 1]$ implies $0 \leqslant x \leqslant 1 < 1 + 1/n$, for all $n \in N$, i.e. $x \in A_n$ for all $n \in N$, i.e. $x \in \cap A_n$. Conversely, $x \in \cap A_n$ implies $0 \leqslant x < 1 + 1/n$, for all $n \in N$, whence $0 \leqslant x \leqslant 1$ (either letting $n \to \infty$, or supposing $x > 1$ and obtaining a contradiction to $x < 1 + 1/n$ for all $n \in N$).

Cover for a set

Let $\mathscr{S}$ be a class of sets A. Then the class $\mathscr{S}$ is called a cover for a set X if and only if

$$X \subset \cup \{A \,|\, A \in \mathscr{S}\}.$$

Any subclass of $\mathscr{S}$ which also covers X is called a subcover of $\mathscr{S}$.

The notion of 'open' cover will be employed in chapter 2, in connection with compact sets. The 'open' here refers to the fact that the sets of the cover are open sets, in the sense of topology. For the moment we shall be content with a very simple example on covers.

Example 2. (i) Let I_n be the open interval

$$(n, n+1) = \{x \in R \,|\, n < x < n+1\}$$

on the real line. Then the class $\{I_n \,|\, n \in Z\}$ is not a cover for R, for no integer belongs to $\cup \{I_n \,|\, n \in Z\}$.

(ii) If $J_n = \{x \in R \,|\, n \leqslant x < n+1\} = [n, n+1)$, then the class $\{J_n \,|\, n \in Z\}$ is a cover for R.

(iii) Let $S[a, r] = \{z \in C \,|\, |z - a| \leqslant r\}$, where $a \in C$ and $r > 0$. Thus $S[a, r]$ is the closed disc of centre a and radius r in the complex plane. It is clear that the class $\{S[m + in, 1] \,|\, m, n \in Z\}$ is a cover for C.

Complementation

If X is our universe of discourse and $A, B \subset X$ then we define

$$A \sim B = \{x \in X \,|\, x \in A, x \notin B\}.$$

We call $A \sim B$ the *complement of B with respect to A*. By $\sim A$ we mean $X \sim A$, and we call $\sim A$ the complement of A. It is clear that $A \sim B = A \cap (\sim B)$, $\sim (\sim A) = A$, and that $A \subset B$ is equivalent to $\sim B \subset \sim A$.

The two following results concerning complementation are known as De Morgan's laws:

$$\sim \cup A_\alpha = \cap (\sim A_\alpha); \quad \sim \cap A_\alpha = \cup (\sim A_\alpha).$$

To prove the first of these, for example, we merely note that $x \in \sim A_\alpha$ for all α is equivalent to $x \notin A_\alpha$ for any α.

Some other properties of union and intersection which are easy to show are

(i) $\cap A_\alpha \subset A_\alpha \subset \cup A_\alpha$, for any α,

(ii) $A \cup (\cap A_\alpha) = \cap (A \cup A_\alpha)$,

(iii) $A \cap (\cup A_\alpha) = \cup (A \cap A_\alpha)$.

Ordered pair

Let x, y be any objects. Then the ordered pair (x, y) is defined as the set $\{\{x\}, \{x, y\}\}$. It is easy to check the fundamental property of ordered pairs: $(x, y) = (u, v)$ if and only if $x = u$ and $y = v$. More generally we may define in a similar way an ordered n-tuple $(x_1, ..., x_n)$ with the property $(x_1, ..., x_n) = (y_1, ..., y_n)$ if and only if $x_1 = y_1, ..., x_n = y_n$.

Relation

A relation ρ is defined to be a set of ordered pairs. For example, $\rho = \{(1, 2), (a, b)\}$ is a relation.

Equivalent notation for $(x, y) \in \rho$ is $x \rho y$. Thus in our example we might write $1\rho 2$ instead of $(1, 2) \in \rho$.

An important type of relation is the

Cartesian product

Let X, Y be given sets. Then

$$X \times Y = \{(x, y) \,|\, x \in X \text{ and } y \in Y\}$$

is called the Cartesian product of X and Y.

Another important relation is the

Equivalence relation

Let X be a given set. A relation ρ is called an equivalence relation on X if and only if it is (i) Reflexive, i.e. $x \rho x$ for all $x \in X$, (ii) Symmetric, i.e. $x \rho y$ implies $y \rho x$, (iii) Transitive, i.e. $x \rho y$ and $y \rho z$ imply $x \rho z$. It is usual to denote an equivalence relation by $\sim$ rather than ρ. There is little danger of confusion with complementation.

Example 3. (i) Equality is obviously an equivalence relation on any set.

(ii) Let $X = \{(x, y) \,|\, x, y \in N\}$. Define $(x, y) \sim (u, v)$ to mean $xv = yu$. Then $\sim$ is an equivalence relation on X. For example, let us check transitivity: $(x, y) \sim (u, v)$ and $(u, v) \sim (z, w)$ imply $xv = yu$, $uw = vz$, whence $xvuw = yuvz$, and so $xw = yz$, i.e. $(x, y) \sim (z, w)$.

(iii) Define $x \sim y$ to mean $x - y$ is divisible by 2. It is easy to check that $\sim$ is an equivalence relation on Z.

Let us return to general relations. The *domain* of a relation is the set of all first co-ordinates of its members. The *range* is the set of all second co-ordinates. If $\sim$ is an equivalence relation on X, then we define $E_x = \{y \in X \mid y \sim x\}$ and call E_x the equivalence class containing the element x. For example, in example 3 (iii) we have

$$E_0 = \{0, \pm 2, \pm 4, \ldots\}.$$

In general, it is easy to check that $E_x = E_y$ if and only if $x \sim y$, and that $E_x \cap E_y = \varnothing$ if $E_x \neq E_y$. It is thus evident that $\{E_x \mid x \in X\}$ forms a partition of X, i.e. X is the union of the disjoint classes E_x. For example, in example 3 (iii) we have

$$Z = \{0, \pm 2, \pm 4, \ldots\} \cup \{\pm 1, \pm 3, \pm 5, \ldots\}.$$

Probably the most significant type of relation that occurs in mathematics is that which is called a function. The following definition of a function may seem rather strange to those who are used to books of analysis which extensively employ functions but never actually define them.

Function

> *A function f is defined to be a relation, such that if $(x, y) \in f$ and $(x, z) \in f$ then $y = z$. Four other terms for function are map, mapping, operator, and transformation.*

Our concept of a function as a certain set of ordered pairs is what some would call the graph of a function, since they define a function as a 'rule' or some such. On occasion we shall use the term 'graph of a function', when this seems more expressive. However, to us, a function and its graph are exactly the same thing.

Example 4. (i) $\{(1, 2), (2, 2)\}$ and $\{(z, z+1) \mid z \in C\}$ are functions.

(ii) $\{(1, 2), (1, 4)\}$ and $\{(x^2, x) \mid x \in R\}$ are not functions. For example $(1, 1)$ and $(1, -1)$ are in the second set.

(iii) $\{(x^2, x) \mid x \in R^+\}$ is a function. Here $R^+ = \{x \in R \mid x > 0\}$. In this case, if the first co-ordinates are equal, $x^2 = y^2$, then $(x - y)(x + y) = 0$, so $x = y$, i.e. the second co-ordinates are equal.

If f is a function and $(x, y) \in f$ then we write $y = f(x)$, which is the conventional notation for y as a function of x. We say that y is the value of f at x, or that y is the image of x under f.

The notation

$$f \colon X \to Y$$

is now widely used in mathematics. It is interpreted as 'f is a function from the set X into the set Y'. The meaning of $f \colon X \to Y$ is that X is the domain of f and that the range of f is a subset of Y, not necessarily the whole of Y.

If $f \colon X \to Y$ and $A \subset X$, then the function $g \colon A \to Y$, defined by $g(a) = f(a)$, for $a \in A$, is called the *restriction* of f to A.

Example 5. (i) Define f by $f(x) = e^x$, for $x \in R$, i.e. the domain of f is R and $f = \{(x, e^x) \,|\, x \in R\}$. The range of f is in fact R^+, as is well-known. We may write, with increasing accuracy, $f \colon R \to R$, and $f \colon R \to R^+$.

(ii) Define f by $f(z) = |z|$, for $z \in C$. Then, with increasing precision, we have $f \colon C \to C$, $f \colon C \to R$, and $f \colon C \to \{x \in R \,|\, x \geqslant 0\}$.

Bijective maps

Let $f \colon X \to Y$. Then f is called injective if $f(x_1) = f(x_2)$ implies $x_1 = x_2$, for every $x_1, x_2 \in X$. If the range of f is the whole of Y, then f is called surjective. A mapping which is both injective and surjective is called bijective.

The terms 'one-one', 'onto' and 'one-one correspondence' are sometimes used instead of 'injective', 'surjective' and 'bijective mapping' respectively.

Example 6. $f \colon R \to \{x \in R \,|\, x \geqslant 0\}$, defined by $f(x) = x^2$ is surjective but not injective. The same prescription for f, but with $f \colon R^+ \to R^+$, is bijective.

Inverse function

Let $f \colon X \to Y$ be bijective. Since f is surjective, if $y \in Y$ then there exists $x \in X$ such that $y = f(x)$. This x is unique, since f is injective. Hence there is an inverse function $g \colon Y \to X$ such that $g(f(x)) = x$, for all $x \in X$, and $f(g(y)) = y$ for all $y \in Y$. It is usual to write $g = f^{-1}$.

Example 7. $f: R \to R^+$, defined by $f(x) = e^x$, is bijective. The inverse $f^{-1}: R^+ \to R$ is denoted by log.

Equivalent sets

Sets X, Y are called equivalent (written $X \sim Y$) if and only if there exists a bijective map $f: X \to Y$. We note that $\sim$ is an equivalence relation. For example, $X \sim X$ since the map $f: X \to X$, given by $f(x) = x$, is a suitable bijection.

Example 8. (i) $\{1, 2, 3\} \sim \{1, 3, 5\}$; $\{2, 4, 6, \ldots\} \sim N$ and $N \sim Z$.

(ii) $\{1, 2\}$ is not equivalent to $\{1, 2, 3\}$.

(iii) The interval $(-1, 1)$ in R is equivalent to R. A suitable bijection is $f(x) = \tan(\pi x / 2)$.

Countable set

A set is called countable if and only if it is equivalent to N or to a subset of N. Otherwise it is called uncountable. In case the set is equivalent to $\{1, 2, \ldots, n\}$ it is called finite, with n elements.

Examples of countable sets are N, Z and Q. The set R of real numbers is uncountable (see Exercise 12). It is clear that any subset of a countable set is countable. Also, if A_n is countable for $n = 1, 2, \ldots$, then $\cup \{A_n | n \in N\}$ is countable, i.e. a countable union of countable sets is countable. To 'count' the elements of $\cup A_n$ we proceed as follows. Let $A_n = \{a_{n1}, a_{n2}, \ldots\}$, $n = 1, 2, \ldots$. Write

$$\{a_{11}, a_{12}, a_{21}, a_{13}, a_{22}, a_{31}, a_{14}, a_{23}, \ldots\}.$$

Then this set is countable and is the union of the A_n, provided we do not allow repetitions. For example, if

$$A_1 = \{2, 3, 4, 5, 6, \ldots\} \quad \text{and} \quad A_2 = \{2, 4, 6, 8, 10, \ldots\}$$

then our scheme gives $A_1 \cup A_2 = \{2, 3, 4, 5, 6, 8, 7, 10, \ldots\}$.

Example 9. Let $A_n = \{m/n | m \in Z\}$, $n = 1, 2, \ldots$. Each A_n is countable, since Z is countable, and $\cup A_n = Q$. Hence Q is countable. Thus Q is equivalent to N, even though N is a very 'thin' subset of Q.

Image and inverse image

Let $f: X \to Y$, and let $A \subset X$, $B \subset Y$. Then the image of the set A under the map f is defined to be $f(A) = \{f(x)|x \in A\}$. The inverse image of B is defined to be $f^{-1}(B) = \{x|x \in X \text{ and } f(x) \in B\}$. Note that $f^{-1}(B)$ is a subset of X; it is not necessary that f should be bijective in order that we may write $f^{-1}(B)$.

The basic properties of image and inverse image are now given.

Theorem 1. *Let $f: X \to Y$, and suppose $\{A_\alpha\}$ is a class of subsets of X and $\{B_\alpha\}$ is a class of subsets of Y. Then*

$$\text{(i)} \quad A_\alpha \subset A_\beta \text{ implies } f(A_\alpha) \subset f(A_\beta);$$

$$\text{(ii)} \quad B_\alpha \subset B_\beta \text{ implies } f^{-1}(B_\alpha) \subset f^{-1}(B_\beta);$$

$$\text{(iii)} \quad f(\cup A_\alpha) = \cup f(A_\alpha);$$

$$\text{(iv)} \quad f(\cap A_\alpha) \subset \cap f(A_\alpha);$$

$$\text{(v)} \quad f^{-1}(\cup B_\alpha) = \cup f^{-1}(B_\alpha);$$

$$\text{(vi)} \quad f^{-1}(\cap B_\alpha) = \cap f^{-1}(B_\alpha);$$

$$\text{(vii)} \quad A_\alpha \subset f^{-1}(f(A_\alpha));$$

$$\text{(viii)} \quad f(f^{-1}(B_\alpha)) \subset B_\alpha.$$

Proof. As a sample we prove (iii). Let $y \in f(\cup A_\alpha)$. Then $y = f(x)$, for some $x \in \cup A_\alpha$, i.e. $y = f(x)$ for $x \in A_\alpha$, some α. Thus $y \in f(A_\alpha) \subset \cup f(A_\alpha)$. Conversely, if $y \in \cup f(A_\alpha)$ then $y \in f(A_{\alpha'})$ for some α', so $y = f(x)$, where $x \in A_{\alpha'} \subset \cup A_\alpha$. Hence $y \in f(\cup A_\alpha)$. This proves (iii). The other results are left as exercises. One should observe the difference between (iv) and (vi).

Composition of functions

Let $f: X \to Y$ and $g: Y \to Z$, where f, g are any functions and X, Y, Z are any sets. Then we define the composite (or product) function $gf: X \to Z$, by $(gf)(x) = g(f(x))$, for each $x \in X$.

A composite function is obviously what is called, in elementary calculus, a function of a function.

Partial-order relation

Let X be a set. Then a partial-order relation on X, denoted in general by $\prec$, is a relation which is: (1) Reflexive (i.e. $x \prec x$ for each $x \in X$). (2) Transitive (i.e. $x \prec y$ and $y \prec z$ imply $x \prec z$). (3) Antisymmetric (i.e. $x \prec y$ and $y \prec x$ imply $x = y$).

A total-order relation, $\prec$, is a partial-order relation with the extra property that for any x, y, either $x \prec y$ or $y \prec x$.

A partially ordered set is just a pair $(X, \prec)$, consisting of a set X and a partial order on it. Similarly for a totally ordered set.

Example 10. (i) Let $\mathscr{A}$ be any collection of sets. Then the set inclusion $\subset$ is a partial order on $\mathscr{A}$. It is not in general a total order on $\mathscr{A}$.

(ii) The 'natural' total order on R, the real numbers, is of course $\leqslant$.

(iii) Define $\prec$ on N by saying that $x \prec y$ if and only if x is an integer multiple of y. Then $\prec$ is obviously a partial ordering of N. It is not total. For example, 3 is not an integer multiple of 2, nor is 2 an integer multiple of 3.

Let $(X, \prec)$ be a partially ordered set. An element $a \in X$ is called maximal if and only if $a \prec x$ for $x \in X$ implies $x = a$. An element $b \in X$ is called minimal if and only if $x \prec b$ for $x \in X$ implies $x = b$.

Suppose $A \subset X$. An element $M \in X$ is called an upper bound for A if and only if $x \prec M$ for all $x \in A$. Similarly, $m \in X$ is a lower bound for A if $m \prec x$ for all $x \in A$.

Example 11. Let X be a non-empty set and let $P(X)$ denote the class of all subsets of X (one calls $P(X)$ the power set of X). Then $P(X)$ is partially ordered by set inclusion. If $\mathscr{A}$ is any class of subsets of X, then $\cup \{A | A \in \mathscr{A}\}$ is an upper bound for $\mathscr{A}$ and $\cap \{A | A \in \mathscr{A}\}$ is a lower bound for $\mathscr{A}$.

In order to prove various important results in several branches of mathematics it has been found necessary to invoke a fundamental axiom of set theory known as Zorn's lemma. We state this now and make some comments on it after.

Zorn's lemma

Let X be a partially ordered set with the property that every totally ordered subset has an upper bound. Then X contains a maximal element.

There are just two places in this book where we need to use this axiom. In chapter 3, section 2, we employ it to prove that every linear space has a Hamel base. Zorn's lemma is again used in the proof of the important Hahn–Banach extension theorem (chapter 4, section 5).

The intuitive meaning of Zorn's lemma does not seem to be immediately apparent, and the reader is asked to accept it as an *axiom* of set theory, which is vital if certain results of an existential type are desired.

We remark that it has been proved that Zorn's lemma is equivalent to two other axioms of set theory: (1) the axiom of choice, (2) the well-ordering principle. For our purposes, Zorn's lemma is, for technical reasons, the best axiom to adopt. Those interested in the equivalence of the three axioms should consult P. C. Suppes' book, which is listed in the bibliography.

Exercises 1

1. Prove that the set equation $(A \cap B) \cup C = A \cap (B \cup C)$ is valid if and only if $C \subset A$.
2. (i) Show that $\{x \in R | e^x = 0\} = \varnothing$.
   (ii) Determine $\{x \in R | e^x = -1\}$ and $\{z \in C | e^z = -1\}$.
3. (i) Let $A_\alpha = (\alpha, \infty)$, the open interval with left-hand end point α, viz. $A_\alpha = \{x \in R | \alpha < x\}$. Prove that
$$\cup \{A_\alpha | \alpha \in R\} = R.$$
   (ii) Let $I_n = (0, 1/n)$. Find $\bigcap\limits_{n=1}^{\infty} I_n$.
4. For any class of sets $\{A_\alpha\}$ and any set A, prove that
$$A \cup (\cap A_\alpha) = \cap (A \cup A_\alpha),$$
$$A \cap (\cup A_\alpha) = \cup (A \cap A_\alpha).$$
5. Prove that $(A \cap B) \times C = (A \times C) \cap (B \times C)$, for any sets A, B, C, where $\times$ denotes the Cartesian product.
6. If (x, y), (u, v) are ordered pairs, prove that $(x, y) = (u, v)$ if and only if $x = u$ and $y = v$.
7. Show that the relation $<$ defines a relation on R, which is transitive but neither reflexive nor symmetric.
8. Let $x = (x_n)$, $y = (y_n)$ be sequences of real numbers. Define $x \sim y$ to mean $x_n - y_n \to 0$. Prove that $\sim$ is an equivalence relation and describe the equivalence class which contains the sequence $(1, 1, 1, \ldots)$.
9. Let $\sim$ be an equivalence relation on a set X. Prove that $E_x = E_y$ if and only if $x \sim y$, and that $E_x \cap E_y = \varnothing$ if $E_x \neq E_y$.
10. Find which of the following relations are functions:
    (i) $\{(2, 1), (1, 2), (\text{Queen}, 1), (\text{Bishop}, 2), (\text{Queen}, 3)\}$,
    (ii) $\{(z, e^z) | z \in C\}$, (iii) $\{(e^z, z) | z \in C\}$.

11. (i) Let $a, b \in R$ and define $f(x) = ax + b$. Prove that $f: R \to R$ is bijective if and only if $a \neq 0$. Illustrate graphically.

(ii) Let $a, b, c \in R$ and define $f(x) = ax^2 + bx + c$ on R. Find the necessary and sufficient conditions for f to be bijective.

12. Let S denote the set of all sequences of 0's and 1's. Thus $s \in S$ means that $s = (s_1, s_2, \ldots)$, where s_n is either 0 or 1. Prove that S is uncountable by assuming otherwise, and 'counting' S as $\{s^{(1)}, s^{(2)}, \ldots\}$, where $s^{(1)} = (s_1^{(1)}, s_2^{(1)}, \ldots)$, $s^{(2)} = (s_1^{(2)}, s_2^{(2)}, \ldots)$, etc. By suitably altering the 'diagonal' elements $s_1^{(1)}, s_2^{(2)}, s_3^{(3)}, \ldots$ it is easy to construct a sequence $s \in S$ such that $s \neq s^{(n)}$ for any $n \in N$. This contradicts the supposed countability of S.

Deduce from the uncountability of S that the set R of all real numbers is uncountable. As a hint think of the elements of R expressed as decimals.

13. Let (a, b), $a < b$, be an open interval in R. Prove that $(a, b) \sim (0, 1)$. Hence show that $(0, 1) \sim R$, in the sense of set equivalence.

14. Find whether or not the set of all irrational numbers in R is countable.

15. Define $f: R \to R$ by $f(x) = x^2$. Find $f((1, 2))$ and $f^{-1}((1, 4))$. Compare with theorem 1 (vii).

16. Prove theorem 1 in its entirety.

17. Prove that $f(A \cap B) = f(A) \cap f(B)$ if and only if f is injective.

18. (i) Let $\mathscr{A} = \{A, \varnothing\}$, where $A \neq \varnothing$. Is $\mathscr{A}$ totally ordered by set inclusion?

(ii) Let $\mathscr{A}$ be the collection of all subsets of the set $\{1, 2\}$. Show that set inclusion is not a total order on $\mathscr{A}$.

2. Real and complex numbers

In this brief section we collect together some of the basic ideas concerning numbers, sequences, limits and series. It is expected that the reader will already be fairly familiar with most of the concepts. All we wish to do is to remind him (or her) of the fundamentals. It will be clear that this section does not provide a course of elementary analysis.

The real number system R may, as is well known, be constructed from the rationals Q, which in turn may be constructed from the positive integers N. Though rewarding, the task is long and arduous, and it is common practice to take a more simple-minded approach and *define* R as a totally ordered field which satisfies the axiom of the upper bound. A totally ordered field is a field, on which is defined a total order relation $\prec$ (see chapter 1, section 1), such that $x \prec y$ implies $x + z \prec y + z$, and also $x \prec y$ and $\theta \prec z$ imply $zx \prec zy$. Here θ denotes the zero element of the field. The rationals Q form a totally ordered field under $\leqslant$, but Q does not satisfy the axiom of the upper bound, which is the essential feature of R.

The axiom of the upper bound for R, sometimes called the completeness axiom (see chapter 2, section 2 in this connection), may be stated as follows: Every nonempty set of real numbers which is bounded above has a real supremum (or least upper bound). In detail, if $S \neq \varnothing$, $S \subset R$ and $s \leqslant H$ for all $s \in S$ and some $H \in R$, then there exists $M \in R$ such that (i) $s \leqslant M$ for all $s \in S$, and (ii) given any $\epsilon > 0$, there exists $s = s(\epsilon) \in S$ such that $s > M - \epsilon$. We write $M = \sup S$, or $M = \sup \{s|s \in S\}$.

From the axiom of the upper bound it readily follows that if $S \neq \varnothing$, $S \subset R$, then S has a greatest lower bound, $\inf S$, whenever S is bounded below.

The whole development of real variable theory, and hence a great part of all analysis, depends crucially on the axiom of the upper bound.

If we now take R for granted we may easily construct the field C of complex numbers. We define C as $R \times R$, with field operations given by

$$(x, y) + (x', y') = (x + x', y + y')$$

and
$$(x, y)(x', y') = (xx' - yy', xy' + yx').$$

We agree to identify $(x, 0)$ with the real number x, since

$$(x, 0) + (x', 0) = (x + x', 0) \quad \text{and} \quad (x, 0)(x', 0) = (xx', 0).$$

Writing $i = (0, 1)$ we thus have $i^2 \equiv i \cdot i = -1$. Hence, if $z = (x, y) \in C$, then $z = x + iy$. It turns out that C cannot be made into a totally ordered field (see exercises 2, question 2).

Now we turn to sequences in C. A complex sequence x is a function $x \colon N \to C$. Thus with each $n \in N$ we associate a complex number $x(n)$; written conventionally as x_n. Also, we write

$$x = (x_n) = (x_1, x_2, \ldots),$$

using round brackets to avoid confusion with a mere set.

Bounded sequence

$x = (x_n)$ is called bounded if and only if there exists $M \geqslant 0$ such that $|x_n| \leqslant M$ for all $n \in N$. The set of all bounded sequences is denoted by l_∞.

Convergent sequence

$x = (x_n)$ is called convergent (with limit l) if and only if, for every $\epsilon > 0$ there exists $N = N(\epsilon)$ such that $|x_n - l| < \epsilon$, for all $n \geqslant N$. We write $x_n \to l$ $(n \to \infty)$, or $\lim x_n = l$, and denote the set of all convergent sequences by c.

Upper and lower limits

If $x \in l_\infty$ is a real sequence, then we define the upper and lower limits of x as

$$\limsup x_n = \inf_{k \geqslant 1}(\sup_{n \geqslant k} x_n),$$

$$\liminf x_n = \sup_{k \geqslant 1}(\inf_{n \geqslant k} x_n).$$

Since x is bounded, $\limsup x_n$ and $\liminf x_n$ are (finite) real numbers. From their definition one finds immediately the following properties $(x, y \in l_\infty)$:

$$\liminf x_n + \liminf y_n \leqslant \liminf(x_n + y_n)$$
$$\leqslant \liminf x_n + \limsup y_n$$
$$\leqslant \limsup (x_n + y_n)$$
$$\leqslant \limsup x_n + \limsup y_n.$$

It is also clear that $x_n \leqslant y_n$, for all n, implies $\limsup x_n \leqslant \limsup y_n$ and $\liminf x_n \leqslant \liminf y_n$.

Cauchy sequence

$x = (x_n)$ is called a Cauchy sequence if and only if

$$|x_n - x_m| \to 0 \ (m, n \to \infty),$$

i.e. for all $\epsilon > 0$, there exists $N = N(\epsilon)$ such that $|x_n - x_m| < \epsilon$, for all $n, m > N$. We denote the set of all Cauchy sequences by $\mathscr{C}$.

The relation between c, $\mathscr{C}$ and l_∞ is given by

Theorem 2. $c = \mathscr{C} \subset l_\infty$, *the inclusion being strict.*

Proof. That $c \subset \mathscr{C} \subset l_\infty$, with $\mathscr{C} \subset l_\infty$ strictly, is trivial. What is noteworthy is that $\mathscr{C} \subset c$. This property, through the properties of $\liminf$ and $\limsup$, stems from the axiom of the upper bound. In fact it can be shown that $\mathscr{C} \subset c$ is *equivalent* to the axiom of the upper bound.

To prove $\mathscr{C} \subset c$, we consider real sequences. The extension to complex sequences is immediate, on observing that

$$|Rl(z)| \leqslant |z| \leqslant |Rl(z)| + |Im(z)|,$$

where $z = x + iy \in C$, $x = Rl(z)$, $y = Im(z)$. Now let $(x_n) \in \mathscr{C}$. Then $|x_n - x_m| < 1$, $n, m \geqslant N(1)$, whence $|x_n| < 1 + |x_N|$, $n \geqslant N$; thus $x \in l_\infty$.

Since x is a real sequence, $\limsup x_n$ and $\liminf x_n$ exist. For all $\epsilon > 0$, there exists N such that

(1) $x_n < x_m + \epsilon$, $n, m \geqslant N$,

(2) $x_n > x_m - \epsilon$, $n, m \geqslant N$.

Fix $m \geqslant N$ and take $\limsup_n$ in (1), $\liminf_n$ in (2), finding

$$\limsup x_n \leqslant x_m + \epsilon, \quad \liminf x_n \geqslant x_m - \epsilon.$$

Hence $\qquad\qquad 0 \leqslant \limsup x_n - \liminf x_n \leqslant 2\epsilon,$

and since ϵ is arbitrary, we have $\limsup x_n = \liminf x_n = l$, say. It follows readily that $x_n \to l$.

The fact that $\mathscr{C} \subset c$ is sometimes called the Cauchy principle for the complex numbers. The idea of Cauchy sequences converging appears in generalized form in chapter 2, section 2, where we discuss complete metric spaces.

As is well known, the theory of infinite series is extremely important in analysis. Here we observe one or two simple things which we will need later, and establish notation. There is considerable confusion as to what an infinite series actually is. Everyone knows that it does not really matter, as long as one makes no blunders in actual manipulations. However, while we have the opportunity *we* shall define a series (of complex terms) to be a pair of sequences (a, s), where $a = (a_n) = (a_1, a_2, \ldots)$ is given, and $s = (s_n)$, is related to a by $s_n = a_1 + a_2 + \ldots + a_n$. Usually we write, as is conventional, Σa_k instead of (a, s) and speak of the series Σa_k. Here our convention is that k runs from 1 to 'infinity'.

By this definition of series, a convergent series ought to be defined as a pair (a, s) such that $s \in c$. Then the set of all convergent series would be $\{(a, s) \,|\, s \in c\}$. However, by well-established convention we define the set γ of all 'convergent series' by

$$\gamma = \{a = (a_k) \,|\, s \in c\}$$
$$= \{a \,|\, \Sigma a_k \text{ converges}\}.$$

By a familiar abuse of language we denote by Σa_k the sum of the series Σa_k, i.e. $\Sigma a_k = \lim s_n$, when $a \in \gamma$. A series Σa_k is called divergent if and only if $a \in \sim \gamma$.

Using theorem 2 we see that $a \in \gamma$ if and only if, for every $\epsilon > 0$, there exists $N = N(\epsilon)$ such that

$$\left| \sum_{k=n}^{n+p} a_k \right| < \epsilon \quad \text{for all } n > N, \text{ and for all } p \geqslant 0.$$

This alternative criterion for $a \in \gamma$ is known as the *general convergence principle* for series. Taking $p = 0$ in it we see that $a \in \gamma$ implies $a_n \to 0$, i.e. $\gamma \subset c_0$, where c_0 denotes the set of all sequences converging to 0.

A series Σa_k is called *absolutely convergent* if and only if $\Sigma |a_k|$ converges. We usually write this as $\Sigma |a_k| < \infty$.

We define the set l_1 of all 'absolutely convergent series' by

$$l_1 = \{a = (a_k) \,|\, \Sigma |a_k| < \infty\}.$$

The general convergence principle shows that $l_1 \subset \gamma$, so that altogether we now have the inclusions $l_1 \subset \gamma \subset c_0 \subset c = \mathscr{C} \subset l_\infty$, all inclusions being strict.

It may be asked why we have used the notation l_∞ for the bounded sequences and l_1 for the absolutely convergent series. The reason is this: define for $0 < p < \infty$,

$$l_p = \{a = (a_k) \,|\, \Sigma |a_k|^p < \infty\}.$$

Then l_1 is the case $p = 1$ of l_p. The set of sequences l_p will prove to be of interest later in this book. The case $p = \infty$ is still somewhat isolated. However, we may observe that $a \in l_p$ is equivalent to $(\Sigma |a_k|^p)^{1/p} < \infty$. It can then be shown that if $a \in l_p$ for some $p > 1$ then

$$\lim_{p \to \infty} (\Sigma |a_k|^p)^{1/p} = \sup |a_k|.$$

But $a \in l_\infty$ is equivalent to $\sup |a_k| < \infty$, so that l_∞ may be regarded as the limiting case of l_p. This should explain the notation l_1 and l_∞, at least with regard to 1 and ∞. The reason for the l has not been considered as only a historian would be interested.

Exercises 2

1. (i) Assuming the axiom of the upper bound for R, prove that every nonempty set of real numbers which has a lower bound has a greatest lower bound.

(ii) Let $x, y \in l_\infty$ be real sequences. Prove that

$$\inf x_n + \inf y_n \leqslant \inf (x_n + y_n) \leqslant \sup (x_n + y_n) \leqslant \sup x_n + \sup y_n$$

and give an example where strict inequality holds throughout.

(iii) Prove that

$$\limsup x_n = \lim_k (\sup_{n \geqslant k} x_n) \quad \text{and} \quad \liminf x_n = \lim_k (\inf_{n \geqslant k} x_n)$$

and deduce that $\liminf x_n \leqslant \limsup x_n$.

(iv) Prove that $\limsup x_n = l$ if and only if, for every $\epsilon > 0$ there exists $N = N(\epsilon)$ such that $x_n < l + \epsilon$ for all $n \geqslant N$, and $x_n > l - \epsilon$ for infinitely many n.

(v) Prove that $x_n \to l$ if and only if $\liminf x_n = \limsup x_n = l$.

2. We know that C is a field. Assume that C is a totally ordered field, under a total order $\prec$. By considering i and 0, under $\prec$, derive a contradiction.

3. (i) Prove that the limit of a convergent sequence is unique.
    (ii) Show that $c \subset \mathscr{C} \subset l_\infty$, and that $\mathscr{C} \subset l_\infty$ is a strict inclusion.
    (iii) Define x by $x_n = 1 + 1 + 1/2! + \ldots + 1/n!$. Use the axiom of the upper bound to show that $x \in c$. Prove the limit of the sequence is not rational.

4. (i) Show that $a \in \gamma$ if and only if $r = (r_n) \in c_0$, where $r_n = \sum\limits_{k=n}^{\infty} a_k$.
    (ii) Show that the inclusions $l_1 \subset \gamma \subset c_0 \subset c \subset l_\infty$ are all strict.
    (iii) Suppose $\limsup |a_n|^{1/n} = l$. Prove that $\Sigma |a_n| < \infty$ if $l < 1$, and Σa_n diverges if $l > 1$. For the case $l = 1$, give examples to show that Σa_n may converge or diverge.

5. Suppose that $\Sigma |a_k|^p < \infty$, for some $p > 0$. Prove that $\Sigma |a_k|^q < \infty$, for $q > p$, and show that
$$\lim_{p \to \infty} (\Sigma |a_k|^p)^{1/p} = \sup |a_k|.$$

3. Sequences of functions, continuity, differentiability

As in section 2, we attempt to survey rapidly the basic concepts mentioned in the title of this section.

Later in the book we shall be generalizing the familiar complex variable concepts of continuity and analyticity of functions. We now recall these, and also quote two well-known theorems from complex variable which will be used in chapter 5.

Let $f : S \to C$, where S is a subset of C. Then f is said to be continuous at $z_0 \in S$ if and only if, for every $\epsilon > 0$, there exists $\delta = \delta(\epsilon, z_0)$ such that $z \in S$, $|z - z_0| < \delta$ imply $|f(z) - f(z_0)| < \epsilon$. This is written as $f(z) \to f(z_0)$ as $z \to z_0$.

It is usual to define analytic functions on domains, i.e. open connected sets in C, rather than on mere sets. Thus $f : D \to C$, D being a domain, is analytic on D if and only if there exists
$$\lim_{h \to 0} \frac{f(z+h) - f(z)}{h} \quad (= f'(z))$$

for every $z \in D$ and $z + h \in D$. It is a familiar, though remarkable, consequence of this definition that the functions $f', f'', \ldots, f^{(n)}, \ldots$ are all themselves analytic on D.

A function which is analytic on the whole plane C is called an integral function. Liouville's theorem states that a bounded integral function must be a constant—a generalized version of this theorem is proved in chapter 4 and used to prove the celebrated Gelfand–Mazur theorem of chapter 5.

Now we quote, for reference, the theorems of Cauchy and Morera (see Ahlfors, 1953).

Cauchy's theorem. Let f be analytic within and on a simple closed path Γ. Then $\int_{\Gamma} f(z)\,dz = 0$.

Morera's theorem. Let f be continuous on a domain D and suppose $\int_{\Gamma} f(z)\,dz = 0$ for every simple closed path Γ in D. Then f is analytic on D.

It frequently happens that a sequence we are examining for convergence depends on a parameter, for example (f_n), where $f_n(z) = z^n$, $z \in C$. In general we consider a sequence $(f_n(s))$ of functions $f_n : S \to C$, $n = 1, 2, \ldots$, where S is an arbitrary set. We say that $f_n \to f \, (n \to \infty)$, pointwise on S, if and only if $f_n(s) \to f(s) \, (n \to \infty)$, for each point $s \in S$. In detail this reads: for each $s \in S$ and for every $\epsilon > 0$, there exists $N = N(\epsilon, s)$ such that $|f_n(s) - f(s)| < \epsilon$, for all $n \geqslant N$. We call f the pointwise limit of $(f_n(s))$ on S.

Example 12. Define $f_n(x) = x^n$ on $S = [0, 1]$ in R. Then f, defined by $f(x) = 0$ on $[0, 1), f(1) = 1$, is the pointwise limit of (x^n) on $[0, 1]$. Note here that although each f_n is continuous on $[0, 1]$, the pointwise limit is discontinuous on $[0, 1]$, though only at one point.

If $f_n \to f \, (n \to \infty)$, pointwise on S, and it happens, as it may, that $N(\epsilon, s)$, which occurs in the detailed definition, is in fact free of s, then we say that $f_n \to f \, (n \to \infty)$, uniformly on S. Explicitly, $f_n \to f \, (n \to \infty)$, uniformly on S, if and only if, for every $\epsilon > 0$, there exists $N = N(\epsilon)$, depending only on ϵ and not on s, such that $|f_n(s) - f(s)| < \epsilon$, for all $n \geqslant N$ and for all $s \in S$.

Example 13. Define $f_n(z) = (z - n)^{-1}$ on the imaginary axis. Then $f_n \to 0 \, (n \to \infty)$, uniformly on this axis. For $|f_n(iy)| \leqslant n^{-1}$, for all $n \geqslant 1$ and for all real y. Hence $|f_n(iy)| < \epsilon$, whenever $n > 1/\epsilon = N(\epsilon)$, for all real y.

It is clear that if $f_n \to f \, (n \to \infty)$, pointwise on S, then $f_n \to f \, (n \to \infty)$, uniformly on S, if and only if

$$\sup_{s \in S} |f_n(s) - f(s)| \to 0 \quad (n \to \infty).$$

The following theorem, which expresses the Cauchy principle for uniform convergence, is often useful.

Theorem 3. *Let $f_n : S \to C$, $n = 1, 2, \ldots$. Then (f_n) is uniformly convergent on S if and only if, for every $\epsilon > 0$, there exists $N = N(\epsilon)$, such that $|f_n(s) - f_m(s)| < \epsilon$, for all $m, n \geqslant N$ and for all $s \in S$.*

Proof. If $f_n \to f$ $(n \to \infty)$, uniformly on S, then it is trivial that (f_n) is a uniform Cauchy sequence, i.e. $|f_n(s) - f_m(s)| \to 0$ $(m, n \to \infty)$, uniformly on S. Conversely, for each $s \in S$, $(f_n(s))$ is convergent by theorem 2. Suppose then that $f_m(s) \to f(s)$ $(m \to \infty)$, for each $s \in S$. Since $\lim_m |f_n(s) - f_m(s)| = |f_n(s) - f(s)|$, for each n, we have $|f_n(s) - f(s)| \leqslant \epsilon/2$ for each $s \in S$ and every $n > N(\epsilon/2)$. Hence

$$\sup_{s \in S} |f_n(s) - f(s)| < \epsilon,$$

for every $n > N(\epsilon/2)$, whence $f_n \to f$ $(n \to \infty)$, uniformly on S.

Two of the main benefits which stem from uniform convergence are given in

Theorem 4. (i) *Let each $f_n : S \to C$ be continuous on $S \subset C$. If $f_n \to f$ $(n \to \infty)$, uniformly on S, then f is continuous on S.*

(ii) *Let Γ be a rectifiable Jordan arc in C, with length l. Suppose $f_n : \Gamma \to C$ is continuous on Γ and that $f_n \to f$ $(n \to \infty)$, uniformly on Γ. Then*

$$\int_\Gamma f_n(z)\, dz \to \int_\Gamma f(z)\, dz \quad (n \to \infty).$$

Proof. (i) Take any $z, z_0 \in S$. By uniform convergence, there exists N such that $|f(z) - f_N(z)| < \epsilon$, $|f_N(z_0) - f(z_0)| < \epsilon$. Since f_N is continuous at z_0, there exists $\delta = \delta(\epsilon, z_0)$ such that $|z - z_0| < \delta$ implies

$$|f_N(z) - f_N(z_0)| < \epsilon.$$

Hence, by the triangle inequality, if $|z - z_0| < \delta$, then $|f(z) - f(z_0)| < 3\epsilon$, so f is continuous at any $z_0 \in S$.

(ii) By part (i), f is continuous on Γ, whence all the integrals which appear exist. Since $|f_n(z) - f(z)| < \epsilon$, for all $z \in \Gamma$ and for all $n \geqslant N$, we have, for such n,

$$\left| \int_\Gamma f_n(z)\, dz - \int_\Gamma f(z)\, dz \right| \leqslant \sup_{z \in \Gamma} \left| f_n(z) - f(z) \right| . l \leqslant \epsilon l,$$

which proves (ii).

With regard to uniform convergence of series of complex-valued functions, we define $\Sigma f_k(s)$ to be uniformly convergent to f on a set S, if and only if each $f_k : S \to C$ and

$$\sum_{k=1}^{n} f_k(s) \to f(s) \quad (n \to \infty), \text{ uniformly on } S.$$

A crude but useful test for uniform convergence of series is

Theorem 5 (Weierstrass M-test). *Let $f_n : S \to C$, $\Sigma M_k < \infty$ and $|f_k(s)| \leqslant M_k$, for all k and for all $s \in S$. Then Σf_k and $\Sigma |f_k|$ are uniformly convergent on S.*

Proof. For $n \geqslant 1$, $p \geqslant 0$, and all $s \in S$,

$$\left| \sum_{n}^{n+p} f_k(s) \right| \leqslant \sum_{n}^{n+p} |f_k(s)| \leqslant \sum_{n}^{n+p} M_k.$$

The result follows from the general convergence principle for series and theorem 3.

Example 14. $\Sigma z^n/n^2$ is uniformly convergent on $|z| \leqslant 1$; for $|z^n/n^2| \leqslant 1/n^2$ on $|z| \leqslant 1$, and $\Sigma 1/n^2 < \infty$.

Exercises 3

1. In example 12, section 3, prove that the convergence of x^n to f is pointwise but not uniform. Prove that the convergence is uniform on $[0, \frac{1}{2}]$.
2. Define $f_n(x) = (1+nx)^{-1}$ on $[0, 1]$. Prove that (f_n) converges pointwise, but not uniformly on $[0, 1]$. Show that $xf_n(x) \to 0$ $(n \to \infty)$, uniformly on $[0, 1]$.
3. Define $f_n(x) = x(1+nx^2)^{-1}$ on R. Prove that $f_n \to 0$ $(n \to \infty)$, uniformly on R. Discuss the validity of the equation $\lim_n f_n'(x) = 0$, for $x \in R$.
4. Let $p > 1$. Prove that $\Sigma z k^{-p}(1+k|z|^2)^{-1}$ converges uniformly on C and that the sum function is continuous on C.
5. $f_n : \{z| \, |z| < 1\} \to C$, $n = 1, 2, \ldots$, and is analytic. Also, $f_n \to f$ $(n \to \infty)$, uniformly on $\{z| \, |z| < 1\}$. Use theorem 4 (ii), Cauchy's theorem, and Morera's theorem to prove that f is analytic on $\{z| \, |z| < 1\}$.
6. Prove that the set of all integral functions f, such that $|f(z)| \leqslant 1$ on C, and $|f(0)| > |f(1)|$, is empty.
7. (Abel's limit theorem.) Let Σa_k be convergent with sum s. Prove that $\Sigma a_k x^k$ converges uniformly on $[0, 1]$ and deduce that $\Sigma a_k x^k \to s$ as $x \to 1-$. As a hint write $s_{n, p} = a_n x^n + \ldots + a_p x^p$, apply Abel's partial summation and obtain $|s_{n, p}| \leqslant \epsilon$, for all $n > N(\epsilon)$, $p > n$ and $x \in [0, 1]$. Then use theorems 3 and 4 (i).

4. Inequalities

We now prove some simple inequalities, which will be freely used in the succeeding chapters.

[1] (The triangle inequality). For any $a, b \in C$, $|a+b| \leqslant |a| + |b|$.

Proof. Since $|Rl(z)| \leqslant |z|$, for any $z \in C$, the result follows on writing $|a+b|^2 = (a+b)(\bar{a}+\bar{b}) \leqslant |a|^2 + |b|^2 + 2|ab|$.

[2] $\dfrac{|a+b|}{1+|a+b|} \leqslant \dfrac{|a|}{1+|a|} + \dfrac{|b|}{1+|b|}$ for any $a, b \in C$.

Proof. Consider $f(t) = t(1+t)^{-1}$, for $t > -1$. We have $f'(t) = (1+t)^{-2}$, for $t > -1$, so that f increases. By the triangle inequality we thus have $f(|a+b|) \leqslant f(|a|+|b|)$, and

$$f(|a|+|b|) = \frac{|a|+|b|}{1+|a|+|b|} \leqslant \frac{|a|}{1+|a|} + \frac{|b|}{1+|b|}.$$

[3] Let $p > 1$, $1/p + 1/q = 1$, $a \geqslant 0$, $b \geqslant 0$. Then $ab \leqslant a^p/p + b^q/q$, with equality if and only if $a^p = b^q$.

Proof. Consider $f(t) = 1 - \lambda + \lambda t - t^\lambda$, where $\lambda = 1/p$ and $t \geqslant 0$. Then $f'(t) < 0$ for $0 < t < 1$ and $f'(t) > 0$ for $t > 1$. Hence $f(t) \geqslant f(1) = 0$, with equality if and only if $t = 1$. Thus we have

$$t^\lambda \leqslant (1-\lambda) + \lambda t, \quad \text{for} \quad t \geqslant 0. \tag{1}$$

If $b = 0$ then $ab = 0 \leqslant a^p/p$. If $b > 0$ we put $t = a^p b^{-q}$ in (1) and obtain our result.

[4] (Hölder's inequality). Let $p > 1$, $1/p + 1/q = 1$, $a_1, \dots, a_n \geqslant 0$ and $b_1, \dots, b_n \geqslant 0$. Then

$$\sum_{k=1}^{n} a_k b_k \leqslant \left(\sum_{k=1}^{n} a_k^p \right)^{1/p} \left(\sum_{k=1}^{n} b_k^q \right)^{1/q}. \tag{2}$$

Also, $$\sum_{k=1}^{n} a_k b_k \leqslant \left(\sum_{k=1}^{n} a_k \right) \cdot \max b_k. \tag{3}$$

Proof. It is (2) which is known as Hölder's inequality. Inequality (3) is trivial and may be regarded as the case $p = 1$ of Hölder's inequality.

To prove (2) let us write

$$A = (\Sigma a_k^p)^{1/p}, \quad B = (\Sigma b_k^q)^{1/q},$$

where the sums run from $k = 1$ to $k = n$. If $AB = 0$ then either $A = 0$ or $B = 0$. In either case we get both sides of (2) equal to zero. Now if $AB > 0$, then by the inequality in [3] above,

$$\frac{a_k}{A} \cdot \frac{b_k}{B} \leqslant \frac{a_k^p}{pA^p} + \frac{b_k^q}{qB^q},$$

whence $\Sigma a_k b_k \leqslant (1/p + 1/q) AB = AB$, which is (2).

It is easy to see from the proof that equality holds in Hölder's inequality if and only if there is a constant M such that

$$a_k^p = Mb_k^q \quad \text{for} \quad 1 \leqslant k \leqslant n.$$

[5] (Minkowski's inequality). Let $p \geqslant 1$, $a_1, ..., a_n \geqslant 0$ and $b_1, ..., b_n \geqslant 0$. Then

$$(\Sigma(a_k + b_k)^p)^{1/p} \leqslant (\Sigma a_k^p)^{1/p} + (\Sigma b_k^p)^{1/p},$$

where the sums run from $k = 1$ to $k = n$.

Proof. The case $p = 1$ is trivial. Suppose $p > 1$. Then omitting the subscript k, for simplicity,

$$\Sigma(a+b)^p = \Sigma a(a+b)^{p-1} + \Sigma b(a+b)^{p-1}$$

$$\leqslant (\Sigma a^p)^{1/p} (\Sigma(a+b)^p)^{1/q} + (\Sigma b^p)^{1/p} (\Sigma(a+b)^p)^{1/q},$$

by Hölder's inequality of [4]. Minkowski's inequality follows immediately.

[6] Let $0 < p \leqslant 1$, $a_1, ..., a_n \geqslant 0$ and $b_1, ..., b_n \geqslant 0$. Then

$$\sum_{k=1}^{n} (a_k + b_k)^p \leqslant \sum_{k=1}^{n} a_k^p + \sum_{k=1}^{n} b_k^p.$$

Proof. It is enough to prove the inequality $(a+b)^p \leqslant a^p + b^p$, valid for $0 < p \leqslant 1$, $a \geqslant 0$, $b \geqslant 0$. To prove this we consider

$$f(t) = 1 + t^p - (1+t)^p,$$

for $t \geqslant 0$. Taking a derivative we find that $f(t) \geqslant 0$, for $t \geqslant 0$, i.e. $(1+t)^p \leqslant 1 + t^p$. If $b = 0$ then $(a+b)^p = a^p + b^p$, and if $b > 0$ we put $t = a/b$ in $(1+t)^p \leqslant 1 + t^p$.

Inequalities [1], [5] and [6] yield the following frequently used results valid for complex a_k, b_k:

$$(\Sigma|a_k + b_k|^p)^{1/p} \leqslant (\Sigma|a_k|^p)^{1/p} + (\Sigma|b_k|^p)^{1/p} \quad (p \geqslant 1),$$
$$\Sigma|a_k + b_k|^p \leqslant \Sigma|a_k|^p + \Sigma|b_k|^p \quad (0 < p \leqslant 1).$$

Exercises 4

1. Let $b \neq 0$. Prove that $|a+b| = |a| + |b|$, if and only if there is a real constant $p \geqslant 0$ such that $a = pb$.

2. Let $p \geqslant 1$. Prove that $\left(\sum_{1}^{n} |a_k| \right)^p \leqslant n^{p-1} \sum_{1}^{n} |a_k|^p$.

3. Let $1 < p < \infty$, $1/p + 1/q = 1$. If $a \in l_p$, $b \in l_q$ prove that $ab \equiv (a_k b_k) \in l_1$. Does $ab \in l_1$ when $a \in l_1$, $b \in l_\infty$?

4. Suppose $1 \leqslant p < \infty$, $a \in l_p$, $b \in l_p$. Write

$$\|a\| = \left(\sum_{k=1}^{\infty} |a_k|^p \right)^{1/p}.$$

Prove that $a+b \equiv (a_k + b_k) \in l_p$ and that $\|a+b\| \leqslant \|a\| + \|b\|$.

5. Suppose $0 < p_k \leqslant 1$, $k = 1, 2, \ldots$ and suppose that $\Sigma |a_k|^{p_k} < \infty$, $\Sigma |b_k|^{p_k} < \infty$. Prove that $\Sigma |a_k + b_k|^{p_k} < \infty$ and that $\Sigma |\lambda a_k|^{p_k} < \infty$, for each fixed $\lambda \in C$.

6. Let $a \in l_p$, $0 < p < 1$. Prove that $\Sigma |a_k| \leqslant (\Sigma |a_k|^p)^{1/p}$.

2

METRIC AND TOPOLOGICAL SPACES

1. Metric and semimetric spaces

In real and complex variable theory there are many results which depend on the algebraic properties of real and complex numbers. For example, the study of power series, so fundamental in complex variable theory, involves the purely algebraic notions of addition and multiplication of complex numbers to form a polynomial $a_0 + a_1 z + \ldots + a_n z^n$. As well as this, one has, of course, to bring in an analytic idea by taking the limit (as $n \to \infty$) of such a polynomial to obtain a power series $\sum_{k=0}^{\infty} a_k z^k$. However, there are many results in elementary analysis which do not depend, in any really essential way, on the algebraic structure of the real or complex numbers. These results involve primarily the idea of distance between numbers x and y.

When we come to generalize the concept of distance so that it applies to the objects in any set we shall not require that the set shall have any algebraic structure, though in particular sets, such as the real or complex numbers, there are 'natural' distance functions employed which do use the algebraic operations. As an example of a basic concept in analysis which is really a metric (or distance) concept we mention the limit l of a function f at a point x_0. The definition of such a limit uses only the distances between x and x_0 and between $f(x)$ and l. From the idea of limit stems the important theory of continuous functions and convergent sequences.

In this chapter we are concerned with some of the basic theory of metric and topological spaces, restricting ourselves to essentials and to those things which will be useful to us in the later chapter on normed spaces.

Metric spaces form the natural generalization of the space of real (or complex) numbers, in so far as it is a space with a distance or metric function. When we have suitably defined a metric space, in a way which most would agree is completely consonant with geometric intuition, we shall have results on limits of functions, continuity, boundedness of sets and convergent sequences, but none, for example, on convergent series. For a series involves addition of elements and in a general metric space there are no algebraic operations. One does of course have metric spaces in which algebraic operations are defined— these are examined in later chapters; for the moment we are concerned with the most general case.

Topological spaces are a natural and important generalization of metric spaces. As we shall see later in the chapter the distance function in a metric space gives rise to a class of sets called open sets and it is the fundamental properties of these open sets which are used to define a topological space.

Now we turn to the precise definition of a

Metric space

A metric space is a pair (X, d), consisting of a nonempty set X and a metric (or distance) function $d : X \times X \to R$ such that, for all x, y, z in X, the following conditions hold,

(M 1) $d(x, y) = 0$ *if and only if* $x = y$,

(M 2) $d(x, y) = d(y, x)$,

(M 3) $d(x, z) \leqslant d(x, y) + d(y, z)$.

Remarks. (i) A metric function is thus a real-valued function defined on pairs of elements of X. It is important to notice that d is necessarily non-negative. For

$$d(x, y) + d(y, x) \geqslant d(x, x)$$

by (M 3), and the other axioms then give $2d(x, y) \geqslant 0$, so that $d(x, y) \geqslant 0$ for all x, y in X.

(ii) The inequality

$$|d(x, y) - d(x', y')| \leqslant d(x, x') + d(y, y'),$$

which follows readily from the axioms, is worth noting.

(iii) Any subset S of a metric space (X, d) can be regarded as a metric space if we restrict d to $S \times S$. Then S is referred to as a subspace of (X, d). When a metric d is fixed on a set X we often loosely refer to X as a metric space, but we should always bear in mind that a metric space is really a pair (X, d), not just a set X.

The axioms (M 1)–(M 3) for a metric space are sometimes referred to as the Hausdorff postulates, after the eminent German mathematician F. Hausdorff (1868–1942).

Axiom (M 3) is commonly called the triangle inequality, since it is the generalization of the geometrically obvious inequality

$$|x-z| \leqslant |x-y| + |y-z|,$$

valid for complex numbers x, y, z.

Now we present several examples of metric spaces.

Example 1. Every nonempty set X can be made into a metric space in a rather trivial way by defining

$$d(x,x) = 0 \quad \text{and} \quad d(x,y) = 1 \quad \text{if} \quad x \neq y.$$

The axioms (M 1)–(M 3) are easily checked. The metric d defined in this way is called the trivial metric on X. One could, for example, put the trivial metric on the set R of real numbers, instead of the usual modulus metric. But then there would be no real variable theory as we know it. The use of the trivial metric is mainly in providing counter-examples to rash assertions and generalizations.

Example 2. Real variable theory is largely the study of R with its 'natural' metric $d(x,y) = |x-y|$. The Hausdorff postulates for a general metric space really enshrine the essential properties of the modulus for real numbers.

The natural metric for the complex numbers C is $d(z,w) = |z-w|$, where $|z|$ is the usual modulus for z in C.

Example 3. Let the positive integers $N = \{1, 2, ...\}$ be augmented by an object ∞ and write $N' = N \cup \{\infty\}$. Define

$$d(m,n) = \left| \frac{1}{m} - \frac{1}{n} \right| \quad \text{for} \quad m, n \in N,$$

$$d(m,\infty) = d(\infty, m) = \frac{1}{m} \quad \text{for} \quad m \in N,$$

$$d(\infty, \infty) = 0.$$

Then it is easy to check that (N', d) is a metric space. Although perhaps a rather curious metric at first sight, d has the advantage that ∞ is actually in N' when one considers such statements as '$n \to \infty$'. This point is further discussed in section 4 below, where we introduce the idea of the limit of a function at a point in a metric space.

Example 4. By R^n, where n is a positive integer, we shall mean n-dimensional Euclidean space, i.e. R^n is the set of all ordered n-tuples $x = (x_1, ..., x_n)$ of real numbers x_i, equipped with the metric

$$d(x, y) = \left(\sum_{i=1}^{n} (x_i - y_i)^2 \right)^{\frac{1}{2}} \quad (x, y \in R^n).$$

The Hausdorff postulates should be checked by the reader—(M 3) follows from Minkowski's inequality of chapter 1.

Note that $R^1 = R$, the real numbers with the natural metric $d(x, y) = ((x_1 - y_1)^2)^{\frac{1}{2}} = |x_1 - y_1|$.

Some other metrics on the set of n-tuples are c and e:

$$c(x, y) = \max \{ |x_i - y_i| \, | 1 \leqslant i \leqslant n \},$$

$$e(x, y) = \sum_{i=1}^{n} |x_i - y_i|.$$

For reference we note the inequalities

$$c(x, y) \leqslant d(x, y) \leqslant \sqrt{n} \cdot c(x, y),$$

$$c(x, y) \leqslant e(x, y) \leqslant n \cdot c(x, y),$$

$$d(x, y) \leqslant e(x, y) \leqslant \sqrt{n} \cdot d(x, y).$$

We may metrize C^n, the set of all n-tuples of complex numbers, in similar ways.

Example 5. (i) Let c be the space of convergent sequences $x = (x_n)$ of complex terms x_n. The natural metric in c is defined by

$$d(x, y) = \sup_n |x_n - y_n|,$$

where $x = (x_n)$, $y = (y_n)$ are sequences in c. The function d is well-defined, since $\sup |x_n - y_n| \leqslant \sup |x_n| + \sup |y_n|$, which is finite because convergent sequences are bounded. (M 1)–(M 2) are easily checked and for every n,

$$|x_n - z_n| \leqslant |x_n - y_n| + |y_n - z_n| \leqslant d(x, y) + d(y, z),$$

whence $\sup |x_n - z_n| \leqslant d(x, y) + d(y, z)$, which is (M 3).

(ii) In c define
$$d(x, y) = |\lim (x_n - y_n)|.$$

Then from the properties of limit and modulus we readily see that (M 2) and (M 3) hold. However, (M 1) is not fully valid, for although $d(x, x) = 0$ we do not have $d(x, y) = 0$ implying $x = y$, e.g. $x_n = 1/n$, $y_n = 0$ for all $n \geqslant 1$. Thus d is not a metric on c, though it is nearly so.

There are many more important examples where this situation arises quite naturally and it prompts the following definition of a space, slightly more general than a metric space.

Semimetric space

A semimetric space (X, d) consists of a nonempty set X and a function $d : X \times X \to R$ called a semimetric, such that $d(x, x) = 0$; $d(x, y) = d(y, x)$; $d(x, z) \leqslant d(x, y) + d(y, z)$ for all x, y, z in X.

The only difference between a metric and a semimetric space is that in the latter space distinct elements can still be zero distance apart. In several situations it is immaterial whether we work in a metric or a semimetric space; usually we deal with metric spaces, since it will be clear whether or not (M 1) is or is not required in some definition or proof.

There is a standard method of turning a semimetric space into a metric space, which we now describe.

Theorem 1. *Let (X, d) be a semimetric space and define $x \sim y$ to mean that $d(x, y) = 0$. Then $\sim$ is an equivalence relation on X and if E_x is the equivalence class containing x; $E = \{E_x | x \in X\}$, then*

$$\rho(E_x, E_y) = d(x, y)$$

is well-defined on $E \times E$ and (E, ρ) is a metric space.

Proof. The proof is simple and is merely indicated. The axioms for a semimetric space imply that $\sim$ is an equivalence relation, e.g. $x \sim y$, $y \sim z$ imply $d(x, y) = d(y, z) = 0$ and so, since d is non-negative, we have $0 \leqslant d(x, z) \leqslant d(x, y) + d(y, z) = 0$, i.e. $d(x, z) = 0$, $x \sim z$. Now we are defining ρ on $E \times E$ by picking x out of E_x and y out of E_y. It must be shown that ρ is independent of which elements are chosen. Now if $x' \in E_x$, $y' \in E_y$ then $x' \sim x$, $y' \sim y$, so that

$$|d(x, y) - d(x', y')| \leqslant d(x, x') + d(y, y')$$

implies $d(x, y) = d(x', y')$. Thus ρ is well-defined. Finally we must check (M 1)–(M 3) and we take (M 1) as a sample. If $E_x = E_y$ then $x \sim y$, $d(x, y) = 0$, $\rho(E_x, E_y) = 0$. Conversely, if $\rho(E_x, E_y) = 0$ then $d(x, y) = 0$, $x \sim y$ and so $E_x = E_y$. We here use the fact that $E_x = E_y$ if and only if $x \sim y$. The check on (M 2)–(M 3) is left as an exercise.

Although the above procedure gives a full metric space, beginning with a semimetric space, it is to be noted that this metric space has rather more complicated elements, viz. equivalence classes, than the original space.

Example 6. Those who may be familiar with the Lebesgue integral should find the following example worthwhile. Let $L = L[0, 1]$ be the set of L-integrable functions f on $[0, 1]$. Then

$$\int_0^1 |f(x)| \, dx < \infty$$

and we may define

$$d(f, g) = \int_0^1 |f(x) - g(x)| \, dx \quad (f, g \in L),$$

which is a semimetric on L. It fails to be a metric since $d(f, g) = 0$ implies only that $f = g$ almost everywhere on $[0, 1]$. By our theorem we then have $f \sim g$ if and only if $d(f, g) = 0$, i.e. if and only if $f = g$ almost everywhere on $[0, 1]$. Thus L may be regarded as a metric space whose elements are equivalence classes of functions, where $g \in E_f$ means that $g = f$ almost everywhere on $[0, 1]$.

Example 7. An important metric space is obtained from the set of continuous real functions $x = x(t)$ on the closed interval $[0, 1]$. Since the modulus of x is continuous, it is bounded and attains its bounds, by a familiar theorem of real variable. Thus we put on the natural metric

$$d(x, y) = \max \{|x(t) - y(t)| \, | \, 0 \leqslant t \leqslant 1\},$$

where x, y are continuous on $[0, 1]$. The metric space obtained is denoted by $C[0, 1]$.

Another metric on the set of continuous functions on $[0, 1]$ is

$$\rho(x, y) = \int_0^1 |x(t) - y(t)| \, dt,$$

where the integral is the ordinary Riemann integral of real analysis. Unlike the Lebesgue case of example 6 we do here have $\rho(x, y) = 0$ implying $x = y$, i.e. $x(t) = y(t)$ for all t in $[0, 1]$. This depends on continuity and amounts to the assertion that $f = f(t)$ continuous and

$$\int_0^1 |f(t)| \, dt = 0$$

imply $f \equiv 0$ on $[0, 1]$. The proof is an exercise.

Remark. In examples 6 and 7 we could of course have $[a, b]$ in place of $[0, 1]$—gaining little except the notation $L[a, b]$, $C[a, b]$.

Example 8. We now give examples of metric sequence spaces, i.e. sets of infinite sequences $x = (x_n)$, x_n complex, which are made into metric spaces in some fairly natural way. As will be seen, the actual method of definition of many of these spaces almost forces us to define the metric in the way we do. Where details of the check on the Hausdorff postulates are not given, the student is advised to supply them.

The space s. By s we denote the space of all possible sequences $x = (x_n)$. Since s is rather an amorphous collection there is no obvious candidate for a metric. The following is most popular:

$$d(x, y) = \Sigma \frac{|x_n - y_n|}{2^n(1 + |x_n - y_n|)} \quad (x, y \in s).$$

The sum runs from 1 to ∞, which will be our convention when we merely write Σ without limits. The factor 2^{-n} ensures convergence of the series, we could have n^{-2} instead. (M 1)–(M 2) are trivial and (M 3) follows from

$$\frac{|a + b|}{1 + |a + b|} \leqslant \frac{|a|}{1 + |a|} + \frac{|b|}{1 + |b|}$$

of chapter 1.

The space l_∞. This is the space of all bounded sequences $x = (x_n)$ with natural metric

$$d(x, y) = \sup_n |x_n - y_n|.$$

The spaces c, c_0. These are subsets of l_∞, both having the l_∞ metric, c being the space of convergent sequences and c_0 the space of null sequences ($x_n \to 0$). In the space c_0 (but not in c) one may actually use $\max |x_n - y_n|$ instead of $\sup |x_n - y_n|$ for the metric. The proof is an easy exercise.

The space $l(p)$. Unlike the spaces above, this space is of fairly recent origin and although reasonably simple it has not been fully explored (see the later chapter on linear metric spaces).

To define $l(p)$ we take a bounded sequence $p = (p_k)$ of strictly positive numbers, so that $0 < p_k \leqslant \sup p_k = H < \infty$. Then

$$l(p) = \{x = (x_k) | \Sigma |x_k|^{p_k} < \infty\}.$$

A natural metric on $l(p)$ is

$$d(x, y) = (\Sigma |x_k - y_k|^{p_k})^{1/M},$$

where $M = \max(1, H)$, as we now show. (M 1)–(M 2) are trivial and for (M 3), by the inequality of chapter 1,

$$|a_k + b_k|^{t_k} \leqslant |a_k|^{t_k} + |b_k|^{t_k}, \tag{$*$}$$

where $t_k = p_k/M \leqslant 1$. Now by $(*)$ and Minkowski's inequality of chapter 1, since $M \geqslant 1$,

$$(\Sigma |a_k + b_k|^{p_k})^{1/M} \leqslant (\Sigma |a_k|^{p_k})^{1/M} + (\Sigma |b_k|^{p_k})^{1/M},$$

so that, with $a_k = x_k - y_k$, $b_k = y_k - z_k$, we have $d(x, z) \leqslant d(x, y) + d(y, z)$, which is (M 3).

The space l_p. This is a special case of $l(p)$ and one should observe the different notation. We define l_p as that case of $l(p)$ in which (p_k) is constant and we write $p_k = p$ for all k, as is conventional. No confusion *ought* to arise between the sequence $p = (p_k)$ of $l(p)$ and the number of p of l_p. Explicitly, for $p > 0$, l_p is the set of all sequences such that $\Sigma |x_k|^p < \infty$. For $p \geqslant 1$ the metric for l_p is, since $M = p$,

$$d(x, y) = (\Sigma |x_k - y_k|^p)^{1/p}.$$

When $0 < p < 1$, since $M = 1$, the metric is

$$d(x, y) = \Sigma |x_k - y_k|^p.$$

The space l_2. This is just the case $p = 2$ of l_p, but is of special importance as we see in chapter 6, since it is the only l_p space which is a Hilbert space.

There are many other examples of metric sequence spaces which are useful in analysis. Some of these will appear in the exercises and later chapters.

Finally, we take an example from complex variable theory.

Example 9. (i) Let A be the set of all complex functions f analytic on $|z| < 1$ and continuous on $|z| \leqslant 1$. By the maximum modulus principle we know that $|f|$ has its maximum on the boundary of $|z| \leqslant 1$, so we define the metric

$$d(f, g) = \max_{|z| \leqslant 1} |f(z) - g(z)| = \max_{|z| = 1} |f(z) - g(z)|,$$

where $f, g \in A$.

(ii) Let I be the set of all integral functions, i.e. functions f which are analytic for all finite z, e.g. polynomials, $\exp z$, $\sin z$, etc. If we write

$$M_n = \max_{|z|=n} |f(z) - g(z)| \quad (n = 1, 2, \ldots),$$

then
$$d(f, g) = \Sigma \frac{M_n}{2^n(1 + M_n)}$$

is a metric on I (cf. the space s above).

The above examples should be sufficient to illustrate the general nature of the metric concept and to show that a very wide range of sets of interest in analysis can be made into metric spaces in a fairly natural way.

Exercises 1

1. In a metric space (X, d) prove that
$$|d(x, y) - d(x', y')| \leqslant d(x, x') + d(y, y').$$
   Deduce that $d(x_n, x) \to 0$ and $d(y_n, y) \to 0$ $(n \to \infty)$ imply
$$d(x_n, y_n) \to d(x, y) \quad (n \to \infty).$$

2. Give a detailed check of the Hausdorff postulates for a trivial metric space (defined in example 1, chapter 2).

3. Let (X_1, d_1), (X_2, d_2) be metric spaces and let $x = (x_1, x_2)$, $y = (y_1, y_2)$ be in the product set $X_1 \times X_2$. Define
$$d(x, y) = \max(d_1(x_1, y_1), d_2(x_2, y_2)).$$
   Find whether $(X_1 \times X_2, d)$ is a metric space.

4. Prove that the space s, of example 8, chapter 2, is a metric space. Show also that
$$\sup d(x, y) = 1,$$
   where the supremum is taken over all elements x, y in s. (Geometrically this says that the space s has diameter 1.)

5. Let X be the set of all sequences $x = (x_n)$. Write $g(z) = \min(1, |z|)$ for any complex z and define
$$d(x, y) = \Sigma \frac{1}{n^2} g(x_n - y_n) \quad (x, y \in X).$$
   Prove that (X, d) is a metric space and that
$$\sup d(x, y) = \tfrac{1}{6}\pi^2,$$
   where the supremum is taken over all x, y in X. Compare exercise 4 above.

6. Verify that the space of continuous functions $C[0, 1]$ of example 7, chapter 2, is a metric space. Do the same for the spaces A and I of example 9, chapter 2.

7. Prove the inequalities stated in example 4, chapter 2.

8. Fill in the details that remain in theorem 1, chapter 2.

9. Let γ be the set of 'convergent series'. Precisely, we mean that $\gamma = \{x = (x_k), x_k \in C | \Sigma x_k \text{ converges}\}$. Prove that (γ, d) is a metric space, with

$$d(x, y) = \sup_n \left| \sum_{k=1}^{n} (x_k - y_k) \right| \quad (x, y \in \gamma).$$

10. Let BV be the set of all $x = (x_n)$ such that

$$\sum_{n=1}^{\infty} |x_n - x_{n+1}| < \infty. \tag{*}$$

An x in BV is called a sequence of bounded variation (BV). Use the defining series condition $(*)$ to define a natural metric on BV. Prove also the set inclusions

$$l_1 \subset BV \subset c$$

and show that BV and γ overlap but neither contains the other. The γ referred to here is that of exercise 9.

11. Let n be a fixed positive integer and let X be the set of all $n \times n$ matrices A with complex entries:

$$A = (a_{ij}) = \begin{pmatrix} a_{11}, \ldots, a_{1n} \\ a_{21}, \ldots, a_{2n} \\ \cdots\cdots\cdots \\ a_{n1}, \ldots, a_{nn} \end{pmatrix}.$$

Define, for A, B in X,

$$d(A, B) = \max\{|a_{ij}| \, | 1 \leqslant i \leqslant n, \, 1 \leqslant j \leqslant n\},$$

$$\rho(A, B) = \max\left\{ \sum_{j=1}^{n} |a_{ij}| \, | 1 \leqslant i \leqslant n \right\}.$$

Prove that (X, d) and (X, ρ) are metric spaces.

2. Complete metric spaces

As well as using the idea of a metric on the real numbers one learns early on in elementary real variable theory that the completeness postulate for the real numbers is of vital importance for the development of the subject. This postulate is usually given in the following form and is called the 'axiom of the upper bound'.

Completeness Axiom 1 for the real numbers

Every nonempty set of real numbers which is bounded above has a real supremum (or least upper bound).

With a view to generalization it is of little use trying to take this axiom so as to define a type of metric space analogous to the reals.

For in a general metric space there is no idea of order between the elements, which is implicit in the above completeness axiom. However, it can be shown that the following postulate (which is purely metric in character) is equivalent to that above.

Completeness Axiom 2 for the real numbers

Every Cauchy sequence (x_n) of real numbers converges to a real number, i.e. if $|x_n - x_m| \to 0$ as $n, m \to \infty$, where $x_n \in R$, $n = 1, 2, \dots$, then there exists $x \in R$ such that $|x_n - x| \to 0$ as $n \to \infty$.

We shall model our definition of a general complete metric space on Axiom 2 for the real numbers. First we must make the obvious definitions of Cauchy sequence and convergent sequence in a metric space (X, d).

Cauchy sequence

A sequence $(x_n) = (x_1, x_2, \dots)$, where $x_n \in X$ for every n, is called a Cauchy sequence in a metric space (X, d) if and only if

$$d(x_n, x_m) \to 0 \quad (m, n \to \infty),$$

i.e. for every $\epsilon > 0$ there exists $N = N(\epsilon)$ such that

$$d(x_n, x_m) < \epsilon \quad \text{for all} \quad n, m > N.$$

The name Cauchy sequence is in honour of the great French mathematician Augustin Louis Cauchy (1789–1857) who may be regarded as the founder of rigorous analysis. Some authors use the term fundamental sequence rather than Cauchy sequence—we give Cauchy the accolade.

Convergent sequence

A sequence (x_n) in (X, d) is called convergent (to x) if and only if there exists $x \in X$ such that $d(x_n, x) \to 0$ $(n \to \infty)$. We then write $x = \lim x_n$ or $x_n \to x$ and call x the limit of the sequence (x_n).

There is an obvious danger associated with the notations $x = \lim x_n$ and $x_n \to x$. They do not, by themselves, make explicit that convergence is taking place in a given metric space (X, d). We emphasize that they cannot be taken in isolation—the context must be examined to see which metric space is involved. If confusion might arise we shall usually say '$x_n \to x$ in the metric d of X' rather than just '$x_n \to x$'.

The obvious way to avoid all danger is to write something like $\lim x_n = x(X, d)$ instead of $\lim x_n = x$, but most mathematicians seem to think that this is unnecessary.

Also we would emphasize that the limit x of (x_n), when it is convergent, must belong to X (see example 10 of this section).

In the definition of convergent sequence we speak of *the* limit of (x_n) rather than *a* limit. This uniqueness is shown in the following proposition.

Theorem 2. (i) *A convergent sequence has a unique limit.*

(ii) *Every convergent sequence is also a Cauchy sequence, but not conversely, in general.*

(iii) *If a Cauchy sequence has a convergent subsequence then the whole sequence is convergent.*

Proof. (i) If $x_n \to x$ and also $x_n \to y$ in the metric d, then by the triangle inequality (M 3),

$$0 \leqslant d(x, y) \leqslant d(x, x_n) + d(x_n, y) \to 0 \quad (n \to \infty).$$

Hence $d(x, y) = 0$, so by (M 1) we have $x = y$, i.e. the limit is unique. Note that if d is merely a semimetric we cannot then conclude from $d(x, y) = 0$ that $x = y$.

(ii) Let (x_n) be convergent, say $x_n \to x$. Then, by (M 1)–(M 3), which we shall invoke in future without reference,

$$0 \leqslant d(x_n, x_m) \leqslant d(x_n, x) + d(x, x_m)$$
$$= d(x_n, x) + d(x_m, x) \to 0 \quad (n, m \to \infty),$$

i.e. $d(x_n, x_m) \to 0 \; (n, m \to \infty)$, so that (x_n) is Cauchy. The 'not conversely' part is shown by an example of a metric space which has a nonconvergent Cauchy sequence.

Example 10. Take $X = (0, 1)$ as a subspace of R, i.e. regard X as a metric space in its own right with the modulus metric. Then $(1/n)$ is Cauchy in X but not convergent in X. For

$$\left| \frac{1}{n} - \frac{1}{m} \right| \leqslant \frac{1}{n} + \frac{1}{m} \to 0 \quad (n, m \to \infty),$$

but $(1/n)$ is trying to converge to 0, which is not in X. More precisely, we know that $1/n \to 0$ in R and that 0 is the unique limit so that no point of $(0, 1)$ could also be the limit.

(iii) Let (x_n) be Cauchy and suppose (x_{n_k}) is a convergent subsequence. By this we mean $(x_{n_k}) = (x_{n_1}, x_{n_2}, \dots)$ where $n_1 < n_2 < n_3 < \dots$ are positive integers and also that $x_{n_k} \to x$ as $k \to \infty$, in the metric d. Thus we have

$$0 \leqslant d(x_n, x) \leqslant d(x_n, x_{n_k}) + d(x_{n_k}, x)$$

and we can make the extreme right member of the inequality as small as we please by taking n and k sufficiently large. Hence $d(x_n, x) \to 0$, so the Cauchy sequence (x_n) converges to the limit of the convergent subsequence (x_{n_k}).

We shall find (iii) of theorem 2 useful in the sequel.

Now with completeness axiom 2 for the real numbers in mind we define a

Complete metric space

A metric space (X, d) is called complete if and only if every Cauchy sequence converges (to a point of X). Explicitly, we require that if $d(x_n, x_m) \to 0$ $(n, m \to \infty)$ then there exists $x \in X$ such that $d(x_n, x) \to 0$ $(n \to \infty)$.

Example 11. The real numbers R with the usual modulus metric form a complete metric space. On the usual *axiomatic* approach to the reals the completeness is merely an axiom, which is axiom 2 above, or, if we take instead axiom 1 then axiom 2 may be deduced from it. If a *constructive* approach to the reals is taken, starting with the rationals Q say, then the completeness of the constructed set has to be proved. Of course Q is not complete with the modulus metric, e.g. the sequence $1 \cdot 4, 1 \cdot 41, 1 \cdot 414, \dots$ is Cauchy in Q but not convergent in Q, since it is 'converging' to $\sqrt{2}$ which is not in Q.

Example 12. Any set X with the trivial metric forms a complete space. For if (x_n) is Cauchy in X then, on taking $\epsilon = \frac{1}{2}$ in the definition of Cauchy sequence, we have $d(x_n, x_{N+1}) < \frac{1}{2}$ for all $n > N = N(\frac{1}{2})$ and so $d(x_n, x_{N+1}) = 0$ for such n, whence it is clear that $x_n \to x_{N+1}$ as $n \to \infty$. Thus every Cauchy sequence converges to a point of X, actually a member of the sequence in this case.

The spaces R^n, c, c_0, $C[0, 1]$, s, l_∞ and $l(p)$ of the earlier examples are all complete metric spaces. The proof of completeness of all these follows a fairly standard pattern, so we take the space $c = c(R)$ of real convergent sequences as a typical example.

Example 13. The space $c = c(R)$ of convergent sequences of real numbers, with $d(x, y) = \sup_i |x_i - y_i|$; $x = (x_i)$, $y = (y_i)$ in c, is complete. For if $(x^{(n)})$ is a Cauchy sequence in c we have

$$d(x^{(n)}, x^{(m)}) \to 0 \quad (m, n \to \infty).$$

Note here that *each member* of the sequence $(x^{(n)})$ is itself a sequence:

$$x^{(n)} = (x_i^{(n)}) = (x_1^{(n)}, x_2^{(n)}, \ldots) \in c \text{ for each } n.$$

Now for each $\epsilon > 0$ there exists N such that $d(x^{(n)}, x^{(m)}) < \epsilon$, $n, m \geqslant N$, i.e. $\sup_i |x_i^{(n)} - x_i^{(m)}| < \epsilon$, $n, m \geqslant N$, and so, *a fortiori*, $|x_i^{(n)} - x_i^{(m)}| < \epsilon$ for $i = 1, 2, \ldots$ and $n, m \geqslant N$. Hence for each i the sequence of real numbers $(x_i^{(m)}) = (x_i^{(1)}, x_i^{(2)}, \ldots)$ is a Cauchy sequence in R, whence by the completeness of R it converges to x_i say, i.e. there exists

$$\lim_m x_i^{(m)} = x_i \quad \text{for each } i.$$

Now fix $n \geqslant N$ and let $m \to \infty$ in $|x_i^{(n)} - x_i^{(m)}| < \epsilon$ to get

$$|x_i^{(n)} - x_i| \leqslant \epsilon \quad \text{for each } i.$$

Note the blunting of the inequality due to taking the limit. Since ϵ is not dependent on i we now have

$$\sup_i |x_i^{(n)} - x_i| \leqslant \epsilon \quad (n \geqslant N), \tag{$*$}$$

which implies that $d(x^{(n)}, x) \to 0$ $(n \to \infty)$, where $x = (x_i)$. This would seem to be the finish, since it appears that $x^{(n)} \to x$ in the metric d. However, we have to go further, since it is possible, *a priori*, that $(x^{(n)})$ is 'convergent' to a sequence x which is not in c and so, on our definition, not convergent at all. The point is that we must show that $x \in c$. It is more convenient to show that x is a Cauchy sequence and once again use the completeness of R. Now the sequence $(x_i^{(N)}) \in c$ and so it is a Cauchy sequence, whence

$$|x_i^{(N)} - x_j^{(N)}| < \epsilon \quad (i, j \geqslant M(\epsilon)).$$

Consequently, for $i, j \geqslant M(\epsilon)$, we have, by $(*)$

$$\begin{aligned} |x_i - x_j| &= |x_i - x_i^{(N)} + x_i^{(N)} - x_j^{(N)} + x_j^{(N)} - x_j| \\ &\leqslant d(x, x^{(N)}) + |x_i^{(N)} - x_j^{(N)}| + d(x^{(N)}, x) \\ &< 3\epsilon, \end{aligned}$$

which proves that x is a Cauchy sequence in R, whence $x \in c$.

Completing an incomplete metric space. When a metric space (X, d) is not complete there is a definite procedure by which it can be completed. This procedure actually uses the completeness of the real line R, so that it cannot be used as it stands to construct R from the incomplete rationals Q. We should like to show how to construct R as a complete metric space from the incomplete space Q. However, the construction is quite arduous and is perhaps slightly out of context in a book of the present nature. For a good account of this constructive approach to the reals, starting from the rationals see L. W. Cohen and G. Ehrlich (1963).

Starting from the completeness of R it is a relatively easy matter to complete a general incomplete metric space (X, d). The reader is referred to exercise 2, question 7 for details of the process of completion.

There is one fact that emerges from the completion of Q to R, which we should mention. It amounts to saying that the rationals are *dense* in the reals. This means that, given any real $\epsilon > 0$ and any real x, there is always a rational q such that $|x - q| < \epsilon$, which is to say that we can get as close as we like to any given real number with a rational number. The concept of dense for general metric spaces is given later in the book.

Exercises 2

1. In a metric space (X, d) prove that the sequence (x_n) converges if and only if every subsequence of (x_n) converges.
2. In (X, d) let the triangle inequality be replaced by the axiom
$$d(x, z) \leq \max(d(x, y), d(y, z)),$$
but make the same definition of Cauchy sequence. Prove that (x_n) is Cauchy if and only if $d(x_n, x_{n+1}) \to 0$ $(n \to \infty)$.
3. Give an example, in R, of a sequence (x_n) such that $|x_n - x_{n+1}| \to 0$ $(n \to \infty)$ but such that (x_n) is not Cauchy (i.e. not convergent, since R is complete). Compare exercise 2 above.
4. Prove that the metric spaces R^n, $C[0, 1]$, s, l_∞ and $l(p)$ of examples 4–8, chapter 2, are complete.
5. A real polynomial $a_0 + a_1 x + \ldots + a_n x^n$ is, as we know, continuous on $[0, 1]$. Regard the set P of all such polynomials (of all degrees n) as a subspace of $C[0, 1]$. Prove that P is not complete. As a hint consider the partial sums of the series for $\exp x$ and observe that $x_n \to x$ in the metric of $C[0, 1]$ means exactly that $x_n(t) \to x(t)$ *uniformly* on $[0, 1]$.
6. The natural metric for $l_1 = \{x \mid \Sigma |x_n| < \infty\}$ is
$$d(x, y) = \Sigma |x_n - y_n|.$$
Since $l_1 \subset l_\infty$ we could put on the 'unnatural' metric
$$\rho(x, y) = \sup_n |x_n - y_n|.$$
Prove that (l_1, d) is complete but that (l_1, ρ) is incomplete.

7. Let (X, d) be incomplete. Use the following hints to prove that there is a complete metric space (Y, ρ) in which X is isometrically embedded as a dense subset. Throughout, $x = (x_n)$, $x' = (x'_n)$ denote sequences in X.

(i) x, x' Cauchy in (X, d) imply $(d(x_n, x'_n))$ converges in R.

(ii) Define $\rho_1(x, x') = \lim_n d(x_n, x'_n)$ on $\mathscr{C}$, the set of all Cauchy sequences in X. Then $(\mathscr{C}, \rho_1)$ is a semimetric space.

(iii) Make $(\mathscr{C}, \rho_1)$ into a metric space (Y, ρ) by using theorem 1, section 1, chapter 2.

(iv) Let $Y_0 \subset Y$ be defined by saying that $E \in Y_0$ if E contains a constant sequence $x_c = (x_1, x_1, \ldots)$. Then Y_0 is isometric to X, i.e. $\rho(E, E') = d(x_1, x'_1)$, where E contains x_c and E' contains x'_c.

(v) Show that $\overline{Y_0} = Y$, i.e. Y_0 is dense in Y (see section 3 for the ideas behind the definition of dense set).

(vi) Use (v) to show that (Y, ρ) is complete. The space (Y, ρ) of equivalence classes of Cauchy sequences from X, with its metric ρ, is called the completion of (X, d).

3. Some metric and topological concepts

In real and complex analysis the idea of neighbourhood of a point is important in many connections. For example, the concept of an open domain in complex variable depends on the notion of neighbourhood. In R a neighbourhood of a point $a \in R$ is an open interval about a, viz. $\{x \mid |x - a| < r\}$ for some $r > 0$. In C a neighbourhood of a point $a \in C$ is an open disc (of radius $r > 0$, say): $\{z \mid |z - a| < r\}$. Noting that only the metric of R (or C) is involved in defining neighbourhood we may define the concept quite naturally in a general metric space:

Neighbourhood (or Open Sphere)

Let $a \in X$, where (X, d) is a metric space. Then, for $r > 0$,

$$S(a, r) = \{x \in X \mid d(a, x) < r\}$$

is a neighbourhood (or open sphere or open ball) of centre a and radius r.

The word 'sphere' originates in ordinary three-dimensional space R^3. In general metric spaces the sphere comes in various nonspherical guises and may even be a single point.

Example 14. (i) $S(a, r)$ is nonempty: $a \in S(a, r)$.

(ii) In C, $S(0, 1)$ is the open unit disc, viz. $\{z \mid |z| < 1\}$.

(iii) In $C[0, 1]$ let θ denote the continuous function which is identically zero on $[0, 1]$. Then $S(\theta, 1)$ is the set of all continuous functions lying (strictly) in a band of width 2 centred on the x-axis.

(iv) Let X be any set with the trivial metric d. If $0 < r \leqslant 1$ then $S(a, r) = \{a\}$, the set consisting of a alone. If $r > 1$ then $S(a, r) = X$, the whole space.

Before we further develop ideas involving neighbourhood we wish to make some definitions which allow us to talk of bounded sets in a metric space.

Distances; diameter

Let (X, d) be a metric space and A, B subsets of X. Define

$$d(x, A) = \inf\{d(x, a) | a \in A\},$$
$$d(A, B) = \inf\{d(a, b) | a \in A, \, b \in B\},$$
$$d(A) = \sup\{d(a, a') | a, a' \in A\}.$$

We call $d(x, A)$ the distance between the point $x \in X$ and the set A, $d(A, B)$ the distance between the sets A and B and $d(A)$ the diameter of the set A.

Bounded Set

A set A in a metric space is called bounded if and only if it has a finite diameter, i.e. $d(A) < \infty$. Otherwise it is unbounded.

Example 15. (i) If A is finite, i.e. has only a finite number of points, then it is obviously bounded and $d(A) = \max\{d(a, a') | a, a' \in A\}$.

(ii) If $A = (0, 1) \subset R$ then $d(A) = 1$, since $|a - a'| < 1$ for all a, $a' \in A$ and for every $\epsilon > 0$ there are obviously points $a, a' \in A$ such that $|a - a'| > 1 - \epsilon$. Note here that the diameter is not attained, i.e. there are no points $a, a' \in A$ such that $|a - a'| = 1$.

(iii) If (x_n) is a Cauchy sequence in (X, d) then $\{x_n\}$ is bounded. For there exists N such that $d(x_n, x_N) < 1$ for all $n > N$ and for the rest of the n we have $M = \max\limits_{1 \leqslant n \leqslant N} d(x_n, x_N) < \infty$, whence $d(x_n, x_N) < M + 1$ for all n, and so for all n and m we have

$$d(x_n, x_m) \leqslant d(x_n, x_N) + d(x_m, x_N) < 2M + 2,$$

whence $d(\{x_n\}) \leqslant 2M + 2 < \infty$.

(iv) c_0 is unbounded, since we can find two null sequences whose distance apart can be made arbitrarily large, e.g. $(0, 0, 0, \ldots)$ and $(n, 0, 0, \ldots)$ for sufficiently large n.

(v) Any metric space (X, d), bounded or not, can be made into a bounded metric space (X, ρ), where

$$\rho(x, y) = \frac{d(x, y)}{1 + d(x, y)}.$$

It is easily checked that ρ is a metric when d is—the triangle inequality being a consequence of inequality [2] of section 4, chapter 1.

An easier example is that in which ρ is the trivial metric on X. Of course, the earlier ρ is related to d. In fact it is easy to see that a sequence in X is convergent (or Cauchy) with respect to d if and only if it is convergent (or Cauchy) with respect to $\rho = d/(1 + d)$.

Now we turn to the concept of open set in a metric space, which is important in connection with continuous functions and also because it gives rise to further generalization to topological spaces. We label open sets by G, from the German *Gebiet* or region. Closed sets (to be defined later) are labelled F, from the French *fermé* or closed.

Open set

> Let (X, d) be a metric space. Then $G \subset X$ is called open if and only if every point of G has a neighbourhood contained in G. Symbolically, we require that if $x \in G$ then there exists $r > 0$ such that $S(x, r) \subset G$.

In the complex plane, which most find is a particularly suitable metric space for visualization, the definition of open set is in reasonable agreement with spatial intuition.

For a general metric space, the fact that an open sphere is so called correctly suggests that it is an open set:

Theorem 3. $S(a, r)$, where $a \in (X, d)$ and $r > 0$, is open.

Proof. Let $x \in S(a, r)$, so that $d(x, a) < r$. Put $r' = r - d(x, a) > 0$. It follows that $S(x, r') \subset S(a, r)$, for $y \in S(x, r')$ implies

$$d(y, a) \leqslant d(y, x) + d(x, a) < r' + d(x, a) = r, \quad \text{i.e.} \quad y \in S(a, r).$$

Thus for each point x in $S(a, r)$ there is a sphere $S(x, r')$ contained in $S(a, r)$.

Part of the next result shows that certain set theoretic combinations of open sets are also open sets. This result is particularly important in that it later provides us with a model for a definition of a general topological space.

Theorem 4. *Let (X, d) be a metric space. Then*

(i) $\varnothing$ *and X are open.*

(ii) *The* union *of any collection (countable or uncountable) of open sets is open.*

(iii) *The* intersection *of a* finite *number of open sets is open.*

Proof. (i) That $\varnothing$ is open may seem slightly odd, since $\varnothing$ has no elements. Some authors just *define* $\varnothing$ to be open to make life easier. We offer a proof by contradiction which we hope will be acceptable. Suppose then that $\varnothing$ is not open. Then the statement 'every element of $\varnothing$ has a sphere about it contained in $\varnothing$' is *false*. This implies a statement which begins '*there exists an element of* $\varnothing$ *such that...*'. The three dots indicate that we have no interest in the rest of the statement. We thus have the contradiction we require: 'there exists an element of $\varnothing$' contradicts the fact that $\varnothing$ has no elements. Hence $\varnothing$ must be open.

X is open since $x \in X$ implies e.g. $S(x, 1) \subset X$.

(ii) If $x \in \cup G_\alpha$ (α running through some indexing set) then $x \in G_\alpha$ for some α and so there exists $S(x) \subset G_\alpha$. We write $S(x)$ for $S(x, r)$ when r is of no interest. Hence $S(x) \subset G_\alpha \subset \cup G_\alpha$, so x in the union implies there exists $S(x)$ in the union, whence the union is open.

(iii) It is enough to do this for *two* open sets—then use induction. Let G_1, G_2 be open. If $I = G_1 \cap G_2$ is empty use (i). When $I \neq \varnothing$ take $x \in I$, so $x \in G_1$ implies $S(x, r_1) \subset G_1$ and also $S(x, r_2) \subset G_2$ for some r_1, r_2. Clearly $S(x, r) \subset I$, where $r = \min(r_1, r_2)$.

Example 16. (i) $\cap \{(-1/i, 1/i) \mid i = 1, 2, ...\} = \{0\}$ in R, so *countable* intersections are not necessarily open (cf. theorem 4 (iii)) since $\{0\}$ is not open.

(ii) In R there are many sets which are not open, e.g. $\{x\}$, $[0, 1]$, $(0, 1]$.

(iii) In contrast with (ii), if (X, d) is a trivial space, then every subset of X is open. For if $A \subset X$, $a \in A$, then $S(a, 1) = \{a\} \subset A$, i.e. A is open.

The interval $[0, 1]$ of (ii) above is called a closed interval in elementary analysis. It is not an open set but this does not mean that 'closed' is to be interpreted as 'not open'. We now wish to use the word closed to describe certain sets in general metric spaces, not merely intervals on the line. Observing that $\sim [0, 1] = (-\infty, 0) \cup (1, \infty)$, a union of open sets (and so open by theorem 4 (ii)), we see that the *complement* of the closed interval $[0, 1]$ is open. This suggests the definition of

Closed set

> *A set in (X, d) is called closed if and only if its complement is open.*

By this definition and De Morgan's laws for complementation we have

Theorem 5. (i) X *and* $\varnothing$ *are closed*, (ii) *intersections of closed sets are closed*, (iii) *finite unions of closed sets are closed*.

Example 17. (i) Every $A \subset X$, X trivial, is closed. For $\sim A \subset X$ and so $\sim A$ is open by example 16 (iii), above. Thus every subset of a trivial metric space is both open and closed.

(ii) $[0, 1)$ is neither open nor closed in R.

(iii) In any metric space the 'closed sphere' $\{x \mid d(x, a) \leqslant r\}$ is a closed set, for it is easy to check that $\{x \mid d(x, a) > r\}$ is open.

The concept of closed set is closely connected with the idea of *limit point*. Let S be a set in (X, d) and x a point of X, not necessarily a point of S. Then x is called a limit point of S if and only if every neighbourhood of x contains a point of S different from x.

The set of limit points of S is denoted by S'.

Example 18. (i) Let $S = (0, 1) \subset R$. Then 1 is a limit point of S, but $1 \notin S$. Also, it is easy to see that $S' = [0, 1]$.

(ii) Let $S = [0, 1] \cup \{2\}$ so that, being a union of two closed sets, S is closed (theorem 5 (iii)). One sees that $S' = [0, 1]$, so that $S \supset S'$. This last property is true for closed sets in general. More precisely we have

Theorem 6. *A set F in a metric space is closed if and only if it contains its limit points, i.e. $F \supset F'$.*

Proof. (i) Let F be closed, so that $\sim F$ is open. If $F' = \varnothing$ then $F \supset F'$. If $F' \neq \varnothing$ take $x \in F'$. Then $x \in F$, for if not then $x \in \sim F$ and so there exists $S(x) \subset \sim F$, since $\sim F$ is open. But $S(x) \subset \sim F$ contradicts the fact that $x \in F'$.

(ii) Let $F \supset F'$. We show that $\sim F$ is open. If $\sim F = \emptyset$ then $\sim F$ is open; otherwise take $x \in \sim F \subset \sim F'$. Thus x is not a limit point of F, whence there exists $S(x) \subset \sim F$. Consequently $\sim F$ is open and F is closed.

We next introduce two further terms which, as we shall see, give alternative ways of describing open and closed sets.

Interior of a set

Let S be any subset of (X, d). The interior $S°$ of S is the largest open set contained in S:

$$S° = \bigcup_{G \subset S} G,$$

i.e. the interior of S is the union of all open sets contained in S.

Closure of a set

Let $S \subset (X, d)$. The closure $\bar{S}$ of S is the smallest closed set containing S:

$$\bar{S} = \bigcap_{F \supset S} F,$$

i.e. the closure of S is the intersection of all closed sets containing S.

There is of course always an open set contained in S and a closed set containing S, viz. $\emptyset$ and X respectively.

The following result is easily proved and is left as an exercise.

Theorem 7. (i) *For any S we have $S° \subset S \subset \bar{S}$.*

(ii) *S is open if and only if $S = S°$.*

(iii) *S is closed if and only if $S = \bar{S}$.*

(iv) *$\bar{S} = S \cup S'$.*

(v) *$S \subset T$ implies $S° \subset T°$ and $\bar{S} \subset \bar{T}$.*

A condition for completeness of a metric space may be given in terms of nested closed spheres. For closed intervals in R the idea is probably already familiar. First we make a definition.

Nest of closed spheres

A nest of closed spheres is a sequence (S_n) of closed spheres S_n such that $S_1 \supset S_2 \supset \ldots$ and such that the diameter $d(S_n) \to 0$ $(n \to \infty)$.

If the space X is complete (no holes!) we would expect every nest of closed spheres to shrink down to a single point, so that $\cap S_n \neq \varnothing$, where the intersection is taken over $n = 1, 2, \ldots$. Precisely, we have

Theorem 8. *Let (X, d) be a metric space. Then (X, d) is complete if and only if every nest of closed spheres has a nonempty intersection.*

Proof. We shall not labour the details. The reader should check each statement. First, let X be complete. Take any nest (S_n). Then if x_n is the centre of S_n we have that (x_n) is Cauchy in X. Hence $x_n \to x \in X$. Also, $x \in \cap S_n$. For, take any n. Then $x_n, x_{n+1}, \ldots$ are in S_n, whence $x \in \bar{S}_n$,‡ where $\bar{S}_n$ is the closure of S_n. But $S_n = \bar{S}_n$ since S_n is closed.

Secondly, let every nest have nonempty intersection. Take any Cauchy sequence (x_n). Determine n_1 such that $d(x_n, x_{n_1}) < \frac{1}{2}$, $n > n_1$. Let $S_1 = S[x_{n_1}, 1] = \{x \mid d(x, x_{n_1}) \leqslant 1\}$. Determine $n_2 > n_1$ such that $d(x_n, x_{n_2}) < \frac{1}{4}$, $n > n_2$. Let $S_2 = S[x_{n_2}, \frac{1}{2}]$. Then $S_2 \subset S_1$. Thus we obtain $n_1 < n_2 < n_3 < \ldots$ and $S_1 \supset S_2 \supset S_3 \supset \ldots$ with $d(S_n) \to 0$. Hence $\cap S_n \neq \varnothing$; actually $\cap S_n = \{x\}$, a single point. Clearly $x_{n_k} \to x$ as $k \to \infty$, so (x_n) has a convergent subsequence. By theorem 2 (iii) we have $x_n \to x (n \to \infty)$, i.e. (x_n) converges, so X is complete.

The following terms, which depend on the idea of closure, will be used in the sequel.

Dense set

A subset S of a metric space (X, d) is called dense in X if and only if $\bar{S} = X$.

Nowhere dense set

$S \subset X$ is called nowhere dense in X if and only if $\bar{S}$ contains no neighbourhood.

Separable space

A metric space (X, d) is called separable if and only if it contains a countable dense subset.

Example 19. (i) Q is dense in R. For $Q \subset R$ and so $\bar{Q} \subset \bar{R} = R$ by Theorem 7 (v), (iii). We must show now that $R \subset \bar{Q}$. Let $x \in R$. Then

‡ See exercise 3, question 13.

for every $\epsilon > 0$ there exists $q \in Q$ such that $|x - q| < \epsilon$, whence $x \in Q$ or $x \in Q'$, so that $x \in Q \cup Q' = \bar{Q}$.

(ii) The integers Z are nowhere dense in R. For Z is closed ($\sim Z$ is a union of open intervals) and so $\bar{Z} = Z$, but it is clear that Z contains no neighbourhood.

Note that $\sim Z$ is dense in R.

(iii) R is separable since Q is a countable dense subset.

Theorem 9. *The following are equivalent*: (i) S *is nowhere dense in* X; (ii) $\bar{S}$ *has empty interior*; (iii) $\sim \bar{S}$ *is dense in* X.

Proof. We prove as a sample that (ii) implies (i), leaving the others as exercises.

Suppose that $\bar{S}^\circ = \varnothing$. Then $\bar{S}$ contains no neighbourhood, for if it did, $\bar{S} \supset S(a)$ say, then $a \in \bar{S}^\circ$, since a belongs to an open set $S(a)$ contained in $\bar{S}$. But $a \in \bar{S}^\circ$ contradicts $\bar{S}^\circ = \varnothing$.

Many of the concepts that we have defined above depend on the idea of open set, e.g. closed set, interior, closure, dense set, etc. An open set itself was defined in terms of neighbourhood which depends strongly on the metric d. However, if we start with a class of sets called 'open', which we require to satisfy certain properties (to be laid down shortly) then we may define closed set, interior, etc., exactly as in a metric space but without reference to a metric. By theorem 4 the class of open sets in a metric space (X, d) is such that $\varnothing$, X are open, any union of open sets is open and so is any finite intersection. We shall take these properties as the *axioms* for a new type of space and then a metric space will be a special case and those results depending only on the use of open set will still hold. The precise definition of this new space is now given.

Topological space

A topological space (X, T) *is a pair consisting of a nonempty set* X *and a class* T *of subsets of* X *satisfying the following axioms*:

(T 1) $\varnothing \in T$ *and* $X \in T$.

(T 2) *Any union (countable or not) of sets in* T *is in* T.

(T 3) *The intersection of any finite number of sets in* T *is in* T.

The sets of T *are called open sets and* T *is called a topology for* X.

The word 'open' should be regarded in a neutral way when one is dealing with a general topological space. Although the word is used in a special sense in a metric space it is in general just a word which describes sets satisfying the axioms (T 1)–(T 3).

Example 20. (i) A metric space is a special kind of topological space, the open sets being defined as before and (T 1)–(T 3) holding by theorem 4.

(ii) Let X be any nonempty set and let T be the set of all subsets of X, including $\varnothing$. Thus we are agreeing to call every subset of X open. That (T 1)–(T 3) hold is almost a matter of language; for example, any union of open sets, i.e. subsets of X, is a subset of X and so is open. This topology is called the *discrete topology* for X, since sets consisting merely of a single point are open.

(iii) In contrast to (ii) the *indiscrete topology* for any set $X \neq \varnothing$ is obtained by taking $T = \{\varnothing, X\}$. Thus in an obvious sense the discrete topology is the finest, and the indiscrete topology the coarsest, topology on X.

(iv) Let X be a three-element set $X = \{x, y, z\}$ and let $T = \{\varnothing, X, \{x\}, \{y\}, \{x, y\}\}$. It is readily checked that (X, T) is a topological space.

(v) Let X be the real line, let T be the set of all open intervals (i.e. intervals of the form (a, b)) and $\varnothing$. Then although the 'open' sets of T are open in the conventional sense, (X, T) is not a topological space, as one sees on examining (T 2).

If d is a metric on X then d gives rise to the class $\mathcal{G}$ of open sets in X and $(X, \mathcal{G})$ is a topological space. Conversely, if a topological space (X, T) is given it is natural to ask whether there exists a metric which gives the topology T.

Metrizable space

A topological space (X, T) is called metrizable if there exists a metric d such that $\mathcal{G} = T$, where $\mathcal{G}$ is the class of open sets determined by d.

Example 21. (i) Let T be the discrete topology on any set X. Then (X, T) is metrizable, for the trivial metric d is such that $\mathcal{G} = T$. This we have essentially proved already in example 16 (iii) above.

(ii) Suppose X has more than one point. Let $T = \{\varnothing, X\}$ be the indiscrete topology on X. Then (X, T) is not metrizable. For suppose it were, i.e. suppose $\mathscr{G} = T$, where $\mathscr{G}$ is the class of open sets determined by the metric d. Then take x, y in X with $x \neq y$ and determine $S(x)$, $S(y)$ such that $S(x) \cap S(y) = \varnothing$ (the radius of both neighbourhoods may be taken as $d(x, y)/2$). But we must have $S(x) = S(y) = X$, since $S(x)$, $S(y)$ are nonempty subsets of $\mathscr{G} = T = \{\varnothing, X\}$. Hence $S(x) \cap S(y) = X = \varnothing$ and this contradiction shows that (X, T) is not metrizable.

In example 21 (ii) we used the fact that, in a metric space, distinct points have disjoint open sets (actually neighbourhoods) about them. A general topological space with this property of separation is called a

Hausdorff space

A topological space (X, T) is called Hausdorff if and only if for any x, y in X with $x \neq y$, there exist two disjoint open sets, one containing x and the other containing y.

There are other kinds of separation axioms which define other types of topological space but is it sufficient for our purposes to know that any metric space is a Hausdorff space. We shall need very little in the way of further topological concepts, the main thing we later introduce is the idea of a compact set (see section 5). In the next section we shall define continuous and semicontinuous functions and related notions on topological spaces. The proofs of many results being no harder for topological spaces than for the more special metric spaces.

Exercises 3

1. Let $R^2 = \{(x_1, x_2) \,|\, x_1, x_2 \in R\}$ have its usual metric. Prove that the distance from the origin to the half-plane $\{x_1 > 0\}$ is zero.
2. Give an example in R of a countable union of closed intervals which is not closed.
3. In a metric space denote the 'closed sphere' of centre a and radius r by $S[a, r] = \{x \,|\, d(x, a) \leqslant r\}$. Prove that

$$\overline{S(a, r)} \subset S[a, r].$$

Give an example to show that the inclusion may be strict.
4. Prove theorem 7, section 3, chapter 2.
5. In R use the fact that bounded monotonic sequences are convergent to prove the nesting principle for closed intervals, i.e. prove that if (I_n) is a nest, $I_n = [a_n, b_n]$, then $\cap I_n \neq \varnothing$. (Cf. theorem 8, section 3, chapter 2.)
6. Prove the remaining implications in theorem 9, section 3, chapter 2.

7. Prove that a closed set is nowhere dense if and only if it contains no open set.

8. Let X be an infinite set (i.e. X has infinitely many points). Let T consist of $\varnothing$, X and all sets G such that $\sim G$ is a finite set. Prove that (X, T) is a topological space.

9. Let $X = [0, 1)$ in R. Let T consist of $\varnothing$ and all sets of the form $[0, a)$ where $0 < a \leqslant 1$. Prove that (X, T) is a topological space.

10. Prove that a metric space is a Hausdorff topological space.

11. Let (X, d) be a metric space and define $\rho = \min(1, d)$. Show that (X, ρ) is a bounded metric space and that d and ρ define the same topology (i.e. $\mathscr{G}$ determined by d is equal to $\mathscr{G}$ determined by ρ).

12. Let (X, T) be a topological space and S a subset of X. Show that S is a topological space if its open sets are defined to be of the form $S \cap G$ where $G \in T$.

This topology on S is called the *induced topology*.

13. Let S be a set in (X, d) and suppose (x_n) is a sequence of points in S such that $x_n \to x$, where $x \in X$. By considering the cases $x \in S$ and $x \notin S$, use theorem 7 (iv) to prove that $x \in \bar{S}$.

4. Continuous functions on metric and topological spaces

In the complex plane C we know that a function $f: C \to C$ is said to have limit l at z_0 if and only if for every $\epsilon > 0$ there exists $\delta = \delta(\epsilon, z_0) > 0$ such that $0 < |z - z_0| < \delta$ implies $|f(z) - l| < \epsilon$. Also, f is said to be continuous at z_0 if and only if $|z - z_0| < \delta$ implies $|f(z) - f(z_0)| < \epsilon$. The difference between the two definitions is that l need not be a value of f in the first definition.

If we have metric spaces (X, d), (Y, ρ) and a function $f: X \to Y$ then we say that f has limit l at $x_0 \in X$ if and only if $0 < d(x, x_0) < \delta$ implies $\rho(f(x), l) < \epsilon$. Similarly, for continuity at x_0, we require $d(x, x_0) < \delta$ implies $\rho(f(x), f(x_0)) < \epsilon$. In the above we have omitted the form of words 'for every $\epsilon > 0$ there exists $\delta = \delta(\epsilon, x_0) > 0$' as we shall usually do in all that follows. When ϵ and δ occur, the form of words will be assumed to have been uttered. Also, in the limit definition it has to be assumed that $l \in Y$.

There is a special type of continuity which is also useful.

Uniform continuity

Let (X, d), (Y, ρ) be metric spaces. Then $f: X \to Y$ is called uniformly continuous on X if and only if for every $\epsilon > 0$ there exists $\delta = \delta(\epsilon) > 0$, depending only on ϵ, such that $d(x, x') < \delta$ implies $\rho(f(x), f(x')) < \epsilon$, where x, x' are in X.

Obviously every uniformly continuous function is also continuous, but not conversely in general. For example, $f(x) = 1/x$ is continuous on $(0, 1)$ but not uniformly continuous, as is easily proved.

The essential feature of uniform continuity is that $\delta(\epsilon)$ depends only on ϵ and not on any particular $x \in X$ as in ordinary continuity. On certain types of set, continuous functions are also uniformly continuous. The idea of compactness will be introduced in section 5 and we shall see that continuity and uniform continuity coincide for compact sets.

Example 21. Let us consider the space (N', d) of example 3, section 1. Now by $f_n \to l$ as $n \to \infty$, in R, we mean $|f_n - l| < \epsilon$ if $n > M(\epsilon)$—we do not mean $|f_n - l| < \epsilon$ if $0 < |n - \infty| < \delta$. This is of course out of line with our general definition of limit in a metric space. In the definition of limit of a convergent sequence (f_n), the symbol ∞ does not appear in the condition for convergence of the sequence (f_n). As far as *meaning* is concerned the symbol ∞ is totally superfluous—it is an expressive fiction which we could well do without in elementary courses of analysis, where many awesome conclusions are drawn from its mystic properties.

However, we can bring $f_n \to l$ as $n \to \infty$ into line with $f(x) \to l$ as $x \to x_0$ if we put the metric d of example 3 on N'. We would then require that, for every $\epsilon > 0$, there exists $\delta = \delta(\epsilon) > 0$ such that $|f_n - l| < \epsilon$ whenever $0 < d(n, \infty) < \delta$. The last inequality would mean $n \in N$ and $d(n, \infty) = 1/n < \delta$, which is equivalent to $n > \delta^{-1}$, or $n > M(\epsilon)$, where $M(\epsilon) = \delta^{-1}$. Thus, with the metric d the convergence of sequences of real numbers has the same formulation as the convergence of functions, and ∞ no longer occupies a privileged place.

The definition of continuity for a metric space is naturally suggested by the situation on the real line. For a topological space it is not so clear how we should define a continuous function. To do this we shall establish an equivalent definition of continuous function on a metric space which involves only the concept of openness. Then we use this to define the general case.

Theorem 10. $f: (X, d) \to (Y, \rho)$ *is continuous on X if and only if the inverse image of every open set in Y is open in X.*

Proof. (i) Let f be continuous on X in the $\epsilon - \delta$ sense. We must show that if $G \subset Y$ is open in Y then $f^{-1}(G)$ is open in X. It may be that

$f^{-1}(G) = \varnothing$, whence $f^{-1}(G)$ is open. If $f^{-1}(G) \neq \varnothing$ take any $x_0 \in f^{-1}(G)$ so that $f(x_0) \in G$. Since G is open there exists $S_Y(f(x_0), \epsilon) \subset G$. By continuity of f there exists $\delta > 0$ such that $x \in S_X(x_0, \delta)$ implies

$$f(x) \in S_Y(f(x_0), \epsilon),$$

i.e. there exists $\delta > 0$ such that $S_X(x_0, \delta) \subset f^{-1}(S_Y(f(x_0), \epsilon))$ and this last set is contained in $f^{-1}(G)$ since f^{-1} is an increasing set function. Hence $x_0 \in f^{-1}(G)$ implies that there exists a neighbourhood

$$S_X(x_0, \delta) \subset f^{-1}(G)$$

so that $f^{-1}(G)$ is open in X.

(ii) For the converse let $f^{-1}(G)$ be open in X for every open G in Y. Take $x_0 \in X$ and any $\epsilon > 0$. Then $f^{-1}(S_Y(f(x_0), \epsilon))$ is open in X and so there exists $S_X(x_0, \delta) \subset f^{-1}(S_Y(f(x_0), \epsilon))$, since x_0 is in this last set. Hence, given any $\epsilon > 0$ there exists $\delta > 0$ such that $x \in S_X(x_0, \delta)$ implies $f(x) \in S_Y(f(x_0), \epsilon)$, i.e. f is continuous in the $\epsilon - \delta$ sense at each point $x_0 \in X$.

Motivated by theorem 10 we now define continuity for functions on topological spaces.

Continuous function on a topological space

Let X, Y be topological spaces. Then $f : X \to Y$ is called continuous on X if and only if the inverse image of every open set in Y is open in X.

Example 22. (i) Let X have the discrete topology and let Y be any topological space. Then every function $f : X \to Y$ is necessarily continuous on X. For $f^{-1}(G)$, where G is open in Y, is a subset of X and so open.

(ii) Let $X = \{x, y, z\}$ and $T = \{\varnothing, X, \{x\}, \{y\}, \{x, y\}\}$ so that (X, T) is a topological space. Define $f : X \to X$ by $f(x) = x$, $f(y) = z$, $f(z) = y$. Then, by considering inverse images of the sets of T, one finds that f is not continuous.

The concept of sequential continuity in a topological space is sometimes useful. If (x_n) is a sequence in a topological space X, then we say that (x_n) converges to $x \in X$ (written $x_n \to x$) if and only if for every open set G containing x there exists $N = N(G)$ such that $n > N$ implies $x_n \in G$.

Let $f : X \to Y$, where X, Y are topological spaces. Then f is called *sequentially continuous* at a point $x \in X$ if and only if, for every sequence $x_n \to x$ (in X) we have $f(x_n) \to f(x)$ (in Y).

Theorem 11. (i) *If* $f : X \to Y$ *is continuous on* X, *then* f *is sequentially continuous on* X, *but not conversely in general.*

(ii) *If* X, Y *are metric spaces, then sequential continuity on* X *implies continuity on* X.

Proof. (i) Take any $x \in X$ and any open set G containing $f(x)$. Then $f^{-1}(G)$ contains x and is open in X. If $x_n \to x$, then $x_n \in f^{-1}(G)$ for almost all n, whence $f(x_n) \in G$ for almost all n, i.e. $f(x_n) \to f(x)$. Thus f is sequentially continuous at each point of X. For the 'not conversely' part see exercises 4, question 14.

(ii) Let (X, d), (Y, ρ) be metric spaces and let f be sequentially continuous on X. Suppose, if possible, that f is not continuous at some point $x \in X$. Then there exists $\epsilon > 0$ and $x_n \in S(x, 1/n)$, for $n = 1, 2, \dots$, such that $\rho(f(x_n), f(x)) > \epsilon$. Hence there is a sequence $x_n \to x$ (in X) such that $f(x_n) \nrightarrow f(x)$ (in Y), which is contrary to the fact that f is sequentially continuous at the point x.

The idea of a semicontinuous function will be used in the sequel and we now briefly describe this object. On the real line the condition for continuity of a real f at x_0 can be split into two: (i) $f(x_0) - \epsilon < f(x)$ if $|x - x_0| < \delta$ and (ii) $f(x) < f(x_0) + \epsilon$ if $|x - x_0| < \delta$. We thus define f as *lower semicontinuous* at x_0 if and only if for every $\epsilon > 0$ there exists $\delta > 0$ such that $f(x_0) - \epsilon < f(x)$ if $|x - x_0| < \delta$. Similarly for *upper semicontinuous* functions. For a topological space we define

Semicontinuous functions

Let X *be a topological space and* $f : X \to R$ *a real function on* X. *Then* f *is called upper semicontinuous on* X *if and only if* $f^{-1}(-\infty, t)$ *is open in* X *for every real* t, *i.e.*

$$\{x \in X \mid f(x) < t\}$$

is open in X *for every real* t. *A function* f *is called lower semicontinuous if* $-f$ *is upper semicontinuous.*

We shall sometimes use the shorter form l.s.c. and u.s.c. to describe the above.

Example 23. (i) If $X = R$ it is easy to check, after the manner of theorem 10, that the above definition is equivalent to the original $\epsilon - \delta$ definition of lower (upper) semicontinuity.

(ii) Let $X = l_\infty$ and define $f: l_\infty \to R \cup \{\infty\}$ by

$$f(x) = \sum_{n=1}^{\infty} |x_n - x_{n+1}|.$$

Then f is lower semicontinuous on l_∞. In this example we are allowing f to take the value ∞ and by convention, if $f(\bar{x}) = \infty$ then f is called l.s.c. at $\bar{x}$ provided there is a neighbourhood of $\bar{x}$ in which $f(x) > \Delta$, for arbitrary $\Delta \in (0, \infty)$. To prove our assertion we consider just the case in which $f(\bar{x}) < \infty$ for $\bar{x} \in l_\infty$. The case $f(\bar{x}) = \infty$ is similar. Now there exists $N = N(\epsilon)$ such that

$$\sum_{1}^{N} |\bar{x}_n - \bar{x}_{n+1}| > f(\bar{x}) - \frac{\epsilon}{2}.$$

Take $\delta = \epsilon/4N$ and let $d(x, \bar{x}) < \delta$, where d is the metric in l_∞. Then

$$\sum_{1}^{N} |\bar{x}_n - \bar{x}_{n+1}| \leqslant 2N d(x, \bar{x}) + \sum_{1}^{N} |x_n - x_{n+1}|,$$

so that

$$f(x) \geqslant \sum_{1}^{N} |x_n - x_{n+1}| > f(\bar{x}) - \epsilon,$$

whence f is lower semicontinuous at any $\bar{x} \in l_\infty$ for which $f(\bar{x}) < \infty$. It is not too difficult to show that f is not continuous on l_∞. In fact one can construct a sequence of elements $x^{(i)} \in l_\infty$ such that $d(x^{(i)}, \theta) = 1/i$ for $i = 1, 2, \ldots$, where $\theta = (0, 0, 0, \ldots)$ and such that $f(x^{(i)}) = 1$ for $i = 1, 2, \ldots$. Thus $x^{(i)} \to \theta$ in the metric of l_∞ but $f(x^{(i)}) \nrightarrow f(\theta) = 0$, so that f is not continuous at θ. The construction of such a sequence $(x^{(i)})$ is left as an exercise.

The next result gives some properties of semicontinuous functions.

Theorem 12. *Let $f: X \to R$, where X is a topological space. Then*

   (i) *f is continuous if and only if it is both l.s.c. and u.s.c.*

   (ii) *f, g l.s.c. (u.s.c.) imply $f + g$ l.s.c. (u.s.c.).*

   (iii) *f_n u.s.c., $f_n \geqslant f_{n+1}$ on X, $f_n \to f$ on X imply f u.s.c.*

   (iv) *f_n u.s.c., $f_n \to f$ uniformly on X imply f u.s.c.*

Proof. We prove (iii) as a sample. From $f_n \geqslant f_{n+1}$ and $f_n \to f$ we infer that $f = \inf f_n$. By hypothesis, for each n, $\{x \mid f_n(x) < t\}$ is open for every real t. Also, we have

$$\{x \mid f(x) < t\} = \bigcup_{n=1}^{\infty} \{x \mid f_n(x) < t\}, \tag{1}$$

for every real t. For if x is in the union in (1) then $f_n(x) < t$ for some n and so $f(x) \leqslant f_n(x) < t$, whence $x \in \{x \mid f(x) < t\}$. Conversely, suppose

$f(x) < t$. Since $f = \inf f_n$ it follows that there exists n such that $f_n(x) < t$, whence x is in the union in (1). Thus, since $\{x \mid f(x) < t\}$ is a union of open sets it is open, whence f is u.s.c.

The rest of the theorem may be proved by similar methods and is left as an exercise.

Earlier we defined $f: X \to Y$ to be continuous if the *inverse* image $f^{-1}(G)$ was open in X when G was open in Y. If a mapping (or function) $f: X \to Y$ has this sort of property for *direct* images it is called an

Open mapping

Let X, Y be topological spaces. Then $f: X \to Y$ is an open mapping if and only if $f(G)$ is open in Y for every open G in X.

Functions which take open sets in one topological space into open sets in another topological space, and conversely, are particularly significant. They are called

Homeomorphisms

Let X, Y be topological spaces. Then $f: X \to Y$ is called a homeomorphism if and only if it is bijective and bicontinuous. The latter means that both f and f^{-1} are continuous.

Equivalently, f is a homeomorphism if and only if it is bijective, continuous and open.

If $f: X \to Y$ is a homeomorphism then X and Y are equivalent as *sets* (f is bijective) and since f and f^{-1} preserve open sets we may regard X and Y as equivalent *topological spaces*, i.e. they may be thought of as indistinguishable from the topological point of view.

We say that two topological spaces are homeomorphic if there exists a homeomorphism between them. To be formal, if we define $X \sim Y$ to mean that X and Y are homeomorphic then it is easy to show that $\sim$ is an equivalence relation on the class of all topological spaces.

Example 24. The open interval $(0, 1)$ and the whole real line R are homeomorphic. A suitable homeomorphism is given by

$$f(x) = \frac{2x - 1}{x(x - 1)}, \quad x \in (0, 1).$$

In fact f is differentiable, not merely continuous on $(0, 1)$ and strictly decreases, whence it is injective. Also, if $y \in R$ then the equation $y = f(x)$ is readily solved for x and one sees that f^{-1} is continuous on R.

Example 25. (R^n, d) and (R^n, c) of example 4, section 1, chapter 2, are homeomorphic. Here a suitable homeomorphism is the identity mapping $f : R^n \to R^n$, given by $f(x) = x$ for each $x = (x_1, \ldots, x_n) \in R^n$. For f is obviously bijective and by example 4, for all x, y in R^n,

$$c(f(x), f(y)) = c(x, y) \leqslant d(x, y) \leqslant \sqrt{n}\, c(x, y).$$

Hence $f(= f^{-1})$ is continuous.

There is a special type of homeomorphism between metric spaces which allows us to regard the spaces as equivalent, provided such a mapping can be found.

Isometry

Let (X, d), (Y, ρ) be metric spaces. Then a mapping $f : X \to Y$ is called an isometry if and only if it is surjective and for all x, x' in X,

$$\rho(f(x), f(x')) = d(x, x').$$

Thus an isometry preserves distances and since it is injective— $f(x) = f(x')$ implies $d(x, x') = 0$, $x = x'$—it makes the sets X, Y equivalent. Clearly f is bicontinuous; in fact uniformly bicontinuous, whence f is a homeomorphism.

Metric spaces are called isometric if there exists an isometry between them. If $X \sim Y$ is taken to mean that X and Y are isometric then $\sim$ is easily seen to be an equivalence relation on the set of all metric spaces.

Example 26. Let R be the metric space of real numbers and R^* the set of all functions f of the form $f(x) = \alpha x$, $\alpha \in R$. If $f(x) = \alpha x$, $g(x) = \beta x$ let us define $d(f, g) = |\alpha - \beta|$. It is easy to check that (R^*, d) is a metric space. Now R and (R^*, d) are isometric under the mapping $T : R^* \to R$ defined by $T(f) = \alpha$, where $f(x) = \alpha x$. For we have $|T(f) - T(g)| = |\alpha - \beta| = d(f, g)$.

This example may seem rather artificial but it is typical of a class of results connected with the dual spaces of normed linear spaces (see chapter 4). In the language of chapter 4 we have just shown that the dual space R^* of R is isometric to R itself.

There is one further type of mapping on a metric space which we shall consider—it has several important applications in connection with existence and uniqueness of solutions of differential and integral equations.

Contraction

> *Let (X, d) be a metric space. Then a mapping $A : X \to X$ is called a contraction if and only if there is a number $c < 1$, independent of x, y in X, such that for all x, y in X,*

$$d(Ax, Ay) \leqslant cd(x, y).$$

We write Ax rather than $A(x)$ for a reason which will shortly appear. This notation is in any case widely used in functional analysis—in multiplicative contexts it is of course liable to abuse.

A contraction A is uniformly continuous on X, for $d(x, y) < \epsilon/c$ implies $d(Ax, Ay) < \epsilon$.

Example 27. Let $f : R \to R$ be differentiable and suppose $|f'(x)| \leqslant c < 1$ on R. Then f is a contraction on R—the mean value theorem gives $|f(x) - f(y)| = |f'(a)| \, |x - y| \leqslant c \, |x - y|$, where $x < a < y$.

Perhaps the most important property of a contraction mapping is its behaviour in a complete space. This is expressed by the *Banach fixed point principle* below. Stefan Banach (1892–1945) was the famous Polish mathematician who was one of the founders of functional analysis.

Theorem 13. *Let the metric space (X, d) be complete and let A be a contraction. Then A has a unique fixed point, i.e. the equation $Ax = x$ has a unique solution for x.*

Proof. Fix on an arbitrary point of X, say x_0 and let $x_1 = Ax_0$, $x_2 = Ax_1 = A^2 x_0, \ldots, x_{n+1} = Ax_n$. We shall show that (x_n) is Cauchy and so convergent in X, to x say. It is this x which is the unique solution of $Ax = x$.

Now for $n \geqslant 1$, $p \geqslant 1$, repeated application of the contraction condition and then the triangle inequality gives

$$
\begin{aligned}
d(x_{n+p}, x_n) = d(A^{n+p} x_0, A^n x_0) &\leqslant cd(x_{n+p-1}, x_{n-1}) \\
&\leqslant c^n d(x_p, x_0) \leqslant c^n(d(x_p, x_{p-1}) + d(x_{p-1}, x_0)) \\
&\leqslant c^n(cd(x_{p-1}, x_{p-2}) + d(x_{p-1}, x_{p-2}) + \ldots) \\
&\leqslant c^n d(x_1, x_0)\,(1 + c + c^2 + \ldots) = c^n d(x_1, x_0)\,(1 - c)^{-1}.
\end{aligned}
$$

Since $c < 1$ we have $d(x_{n+p}, x_n) \to 0$ $(n \to \infty)$, i.e. (x_n) is Cauchy. By completeness of X we have $x_n \to x$, so $x_{n+1} \to x$, and since A is continuous, theorem 11 shows that $Ax_n \to Ax$. Letting $n \to \infty$ in $x_{n+1} = Ax_n$ we then get $x = Ax$.

Although x appears to depend on the starting point x_0 it is in fact the unique solution of $x = Ax$. For if y is another solution then $d(x, y) = d(Ax, Ay) \leqslant cd(x, y)$, which implies $d(x, y) = 0$, whence $x = y$.

The beauty of the method is that one arrives at the unique solution wherever one starts—though one has to admit that the arrival depends on letting $n \to \infty$. In a numerical procedure one must of course make a judicious choice of the starting point.

Example 28. Let f be a real differentiable function on $[a, b]$, with a bounded derivative and suppose $f(x) = 0$ has a solution in $[a, b]$, e.g. $f(a)f(b) < 0$. Provided suitable bounds can be found for f' the contraction method can be applied to $A(x) = x - \lambda f(x)$ where λ is a parameter to be chosen. For if λ can be chosen so as to make A a contraction, say $a \leqslant A(x) \leqslant b$ and $|1 - \lambda f'(x)| \leqslant c < 1$ on $[a, b]$, then $A(x) = x$ will have a unique solution and so $\lambda f(x) = 0$ will have this solution, whence $f(x) = 0$ will have it.

We should note that the above procedure uses the fact that a closed interval $[a, b]$ is a complete metric space in R. This is a special case of a more general result on closed subsets of a complete metric space which is worth giving (slightly out of our present context).

Theorem 14. *A closed subspace F of a complete metric space X is complete.*

Proof. Let (x_n) be Cauchy in $F \subset X$. Then, being Cauchy in X, (x_n) converges, to $x \in X$. But $x \in \bar{F}$, since for every $\epsilon > 0$ there exists $x_N \in F$ such that $d(x_N, x) < \epsilon$. Thus $x \in F = \bar{F}$, since F is closed. Hence (x_n) converges to a point of F, so that F is complete.

There is a result related to theorem 14 to the effect that a complete subspace of a metric space (not necessarily complete itself) is closed. This is left as an exercise.

The following theorem, which has some useful applications, is a corollary to the Banach fixed point principle. It has the virtue that a positive integral parameter n is included, which may be chosen to meet the situation in hand.

Theorem 15. *Let (X, d) be complete and $A : X \to X$. If A^n is a contraction for some $n \geqslant 1$ then $Ax = x$ has a unique solution.*

Proof. Write $B = A^n$, so by theorem 13, $Bx = x$ has a unique solution x. Now $ABx = Ax$ and $AB = BA = A^{n+1}$, whence $BAx = Ax$ and so

$$d(x, Ax) = d(Bx, BAx) \leqslant cd(x, Ax)$$

which implies $x = Ax$. That x is the unique solution of $Ax = x$ is a straightforward exercise.

Note that it is not necessary to assume that A is continuous in theorem 15.

Example 29. The integral equation

$$x(t) = \lambda \int_a^t K(t, u) x(u) \, du + \phi(t) \tag{2}$$

has a unique continuous solution for $x = x(t)$ on $[a, b]$, where λ is an arbitrary parameter, ϕ is continuous on $[a, b]$ and K is continuous on $[a, b] \times [a, b]$. To show this consider $A : C[a, b] \to C[a, b]$ defined by

$$Ax(t) = \lambda \int_a^t K(t, u) x(u) \, du + \phi(t).$$

This is well-defined, since x continuous on $[a, b]$ implies that the integral exists and is a continuous function of the upper limit t. Now for $x, y \in C[a, b]$ an easy induction shows that

$$|A^n x(t) - A^n y(t)| \leqslant |\lambda|^n \, M^n m(t-a)^n / n! \tag{3}$$

on $[a, b]$, where $m = d(x, y) = \max \{|x(t) - y(t)| \, |a \leqslant t \leqslant b\}$ is the metric on the complete space $C[a, b]$ and M is the maximum of $|K(t, u)|$ on the rectangle $[a, b] \times [a, b]$. Since $n!$ outweighs the powers in (3) we see that A^n will be a contraction as soon as n is large enough. Hence $x = Ax$ has a unique solution for x, i.e. (2) has a unique solution $x \in C[a, b]$.

Exercises 4

1. Define f, g, h by $f(x) = 2x, g(x) = x^2, h(x) = 1/x$. Show that f is uniformly continuous on R; g is continuous but not uniformly continuous on R; g is uniformly continuous on $(0, 1)$; h is continuous but not uniformly continuous on $(0, 1)$.
2. Let X be a metric space and $f : N \to X$ any function on the positive integers N. Prove that f is uniformly continuous on N.
3. Let $A \neq \varnothing$ be a given subset of a metric space (X, d). If $f(x) = d(x, A)$ show that f is uniformly continuous on X.

4. Explicitly construct a sequence $(x^{(i)}) \in l_\infty$, as described in the last part of example 23, section 4, chapter 2.

5. Prove (i), (ii) and (iv) of theorem 12, section 4, chapter 2.

6. Let X, Y be discrete topological spaces. Prove that X and Y are homeomorphic if and only if they are equivalent as sets (i.e. cardinally equivalent).

7. Show that $f(x) = 1/x$ defines a homeomorphism $f:(0,\infty) \to (0,\infty)$. Prove that f does not preserve Cauchy sequences in $(0,\infty)$, i.e. show that there is a Cauchy sequence (x_n) in $(0,\infty)$ such that $(f(x_n))$ is not Cauchy in $(0,\infty)$.

8. Give an example of a mapping $f:R \to R$ which is continuous but not open.

9. Prove that any closed interval $[a,b]$ in R, $a < b$, is homeomorphic to $[0,1]$. Is this true if $a = b$?

10. Let X, Y be topological spaces. Prove that $f:X \to Y$ is continuous if and only if the inverse image of every closed set in Y is closed in X.

11. Prove that a complete subspace of a metric space is closed (see theorem 14, section 4, chapter 2).

12. Let n be fixed and define $\rho(x,y) = \max |x_i - y_i|$ on R^n. Let $y = Ax$ be defined by

$$y_i = \sum_{j=1}^{n} a_{ij} x_j + b_i \quad (1 \leqslant i \leqslant n).$$

Prove that A is a contraction mapping on (R^n, ρ) if and only if

$$\max_i \sum_{j=1}^{n} |a_{ij}| < 1.$$

13. Suppose that (a_{ij}), $i,j = 1,2,3,\ldots$, is an infinite matrix. Let

$$\sup_i \sum_{j=1}^{\infty} |a_{ij}| < 1.$$

Prove that the system of equations

$$x_i = \sum_{j=1}^{\infty} a_{ij} x_j + b_i \quad (i = 1,2,3,\ldots),$$

where (b_i) is bounded, has a unique bounded solution.

14. Let T consist of $\varnothing$ and the complements of countable sets in R. Prove that T is a topology for R. Show that a sequence (x_n) converges to x in this topology, if and only if $x_n = x$, for all sufficiently large n.

Prove also that every function $f:(R,T) \to (X,T')$, where (X,T') is any topological space, is sequentially continuous at each point of R. Show that $f(x) = x$ is sequentially continuous but not continuous on R.

15. Let $f:(X,T) \to (Y,T')$ be continuous on X and let $S \subset X$. Prove that the restriction of f to S is continuous with respect to the induced topology for S (see question 12, exercises 3, chapter 2).

5. Compact sets

We consider sets in the general context of a topological space. Some of our results apply especially or only in metric spaces. Our treatment is basic and concerned only with certain results of particular interest— few converses or partial converses of theorems are given.

As a first approximation, a compact set in a topological space may be regarded as a generalization of a closed interval $[a, b]$ on the real line. For example, the elementary theorem that f continuous on $[a, b]$ implies f bounded and attains its bounds still holds in a topological space, provided we replace $[a, b]$ by a compact set K, supposing still that f is real-valued.

The definition of a general compact set is suggested by the following well-known Heine–Borel theorem for the real line (some give Heine no credit and call it Borel's theorem—we shall not concern ourselves with historical niceties).

Theorem 16. *Let F be a bounded closed set in R. Then every open cover of F has a finite subcover.*

Proof. For the definition of cover we refer to section 1, chapter 1. By open cover we mean cover whose sets are open sets. A finite subcover consists of a finite number of sets of the open cover.

Since F is bounded we have $F \subset I_1$, for some closed interval $I_1 = [a, b]$ with $a < b$. Since F is closed it is clear that if every open cover of I_1 contains a finite subcover then every open cover of F will contain a finite subcover. Let us prove the result for I_1. Suppose on the contrary that there is an open cover $\{G_\alpha\}$ of I_1 which has no finite subcover. Bisect I_1 into two closed intervals I_2, I_2'. At least one of these must have no finite subcover. Relabel so that I_2 has no finite subcover. Bisect I_2 and obtain I_3 which has no finite subcover. Continuing in this way we obtain a nest (I_n) of closed intervals, such that for every n, I_n has no finite subcover. By question 5, exercises 3, chapter 2, we have $\cap I_n \neq \varnothing$. Choose $x \in \cap I_n$. Then $x \in I_n \subset I_1 \subset \cup G_\alpha$, so that $x \in G_\alpha$ for some α. Since G_α is open there exists $S(x, r) \subset G_\alpha$. Now by taking n large enough we have $I_n \subset S(x, r) \subset G_\alpha$, whence I_n is covered by the single set G_α. This contradicts the fact that I_n has no finite subcover. The theorem is thus proved.

In view of theorem 16, if we call a set K *compact* if and only if every open cover of K has a finite subcover, then each bounded closed set in R is compact. The converse is in fact also true, i.e. each compact set in R is bounded and closed (see theorem 17 below). The definition of compact involves only the idea of open set and is free of the metric concept of boundedness. Hence in general we define

Compact set

Let X be a topological space. Then a subset K of X is called compact if and only if every open cover of K has a finite subcover. Explicitly, if $\{G_\alpha\}$ is a collection of open sets (countable or not) which covers K, then there exists a finite collection of sets $G_{\alpha_1}, ..., G_{\alpha_n}$ which covers K.

Some properties of compact sets are now given.

Theorem 17. (i) *In a topological space a finite union of compact sets is compact.*

(ii) *A closed subset F of a compact set K is compact.*

(iii) *In a metric space a compact set K is closed and bounded, but not conversely in general.*

Proof. (i) is easy and is left as an exercise.

(ii) To show F is compact let $\cup G_\alpha \supset F$. We must extract a finite subcover. Now $\sim F$ is open since F is closed and so $\cup G_\alpha \cup (\sim F) \supset K$. Since K is compact we have that $\{G_{\alpha_1}, ..., G_{\alpha_n}, \sim F\}$ covers K. This implies that $\{G_{\alpha_1}, ..., G_{\alpha_n}\}$ covers F, i.e. F has a finite subcover.

(iii) We leave the closed part as an exercise. The idea is to show that $\sim K$ is open by taking p in $\sim K$ and proving that p has a neighbourhood contained in $\sim K$.

To show that K is bounded we fix x in K and note that

$$K \subset \bigcup_{i=1}^{\infty} S(x, i).$$

Since K is compact, a finite number of the concentric spheres $S(x, i)$ cover K. Thus K is contained in the largest of these, whence K is bounded.

Finally let (X, d) be a trivial metric space. Then X is closed and bounded ($d(X) = 1$ unless X consists of a single point). However, if X has infinitely many points, e.g. $X = \{1, 2, 3, ...\}$ then X is not compact. For X is the union of the open sets $\{1\}, \{2\}, ...$, but no finite number of these sets can cover X.

We now look at functions on compact sets. The continuous image of an open set need not be open, e.g. $G = (-1, 1), f(x) = x^2$. The same holds for closed sets as is easy to see in the same way. For compact sets the situation is different.

Theorem 18. *The continuous image of a compact set K is compact.*

Proof. Let X, Y be topological spaces and $f : X \to Y$ be continuous on X. By hypothesis $K \subset X$ is compact. Take an open covering $\{G_\alpha\}$ of $f(K)$. Then, using properties of f^{-1} established in chapter 1,

$$K \subset f^{-1}(\cup G_\alpha) = \cup f^{-1}(G_\alpha),$$

whence the same inclusion holds where the union is over a *finite* number of α. Taking the direct image f we have

$$f(K) \subset \cup f(f^{-1}(G_\alpha)) \subset \cup G_\alpha,$$

so there is a finite subcover for $f(K)$. The continuity of f appears when we use the fact that $f^{-1}(G_\alpha)$ is open.

Theorem 19. *A continuous real function on a compact set is bounded and attains its bounds.*

Proof. Let $f : X \to R$, where X is a topological space and R has its usual topology. Let $K \subset X$ be compact. By theorem 18, $f(K) \subset R$ is compact, whence closed and bounded by theorem 17. The boundedness of f is thus established and so there exists $M = \sup \{f(x)|\, x \in K\}$. Hence there exists $x \in K$ such that $M - \epsilon < f(x) \leqslant M < M + \epsilon$, which implies that M is in the closure of $f(K)$. But $f(K)$ is closed and so equal to its closure, whence $M \in f(K)$, i.e. $M = f(x)$ for some $x \in K$. The infimum is attained in a similar way.

This result may be used to metrize the set $C(K)$ of all continuous real functions on a compact set K in a topological space. The natural metric is
$$d(f, g) = \max \{|f(x) - g(x)|\, |x \in K\},$$
where $f, g \in C(K)$.

The next result shows that, in a metric space, compactness is stronger than completeness.

Theorem 20. *If (X, d) is compact then it is complete, but not conversely, in general.*

Proof. Let X be compact and let (S_n) be a sequence of nested closed spheres. Then $\cap S_n \neq \varnothing$. For if the intersection were empty, then, taking complements, we see that one of the spheres would be empty, which it is not. Hence, by theorem 8 we have that X is complete.

The example $X = R$ shows the converse false.

We now wish to give an alternative, more analytical, description of compact metric spaces, which is often useful. Before we do this we shall need two further definitions.

Sequential compactness

A metric space X is said to be sequentially compact if and only if every sequence in X has a convergent subsequence.

Total boundedness

A metric space X is called totally bounded if and only if, for every $\epsilon > 0$ there exists a finite collection of ϵ-spheres which covers X:

$$X = \bigcup_{i=1}^{n} S(a_i, \epsilon); \quad n = n(\epsilon).$$

The set $A = \{a_1, ..., a_n\}$ is then called a finite ϵ-net for X.

It is easy to see that a totally bounded space is bounded but that the converse is false in general (see the exercises).

We shall now show that compactness and sequential compactness amount to the same thing in metric spaces. Some authors, particularly Russian authors, use the term compact to describe what we call sequentially compact. In view of what we are about to prove this usage creates no problems. However, our terminology is standard amongst American and Western European mathematicians.

Theorem 21. *A metric space is compact if and only if it is sequentially compact.*

Proof. (i) First let the space X be compact. By theorem 20 it follows that X is complete. From compactness one immediately deduces total boundedness: cover X with $\{S(x, \epsilon) | x \in X\}$ and extract a finite subcover.

Now let (x_n) be a sequence in X. By total boundedness there is a finite 1-net and so at least one of the spheres of this net contains an infinite subsequence $(x_n^{(1)})$ of (x_n). Let this sphere be S_1, of radius 1.

Also, there is a finite $\frac{1}{2}$-net and so there exists S_2, of radius $\frac{1}{2}$, which contains a subsequence $(x_n^{(2)})$ of $(x_n^{(1)})$. We continue this process and then examine the subsequence $(x_1^{(1)}, x_2^{(2)}, x_3^{(3)}, \ldots)$. This is clearly a Cauchy sequence, for m, $n > N$ implies $x_n^{(n)}$, $x_m^{(m)} \in S_N$. Since X is complete, we have $(x_n^{(n)})$ convergent, so we have shown that X is sequentially compact.

(ii) Now suppose that X is sequentially compact. First we show that X must be totally bounded and then use this to show that X must be compact.

Let $x_1 \in X$ and take $\epsilon > 0$. If $d(x, x_1) < \epsilon$ for all x then $\{x_1\}$ is a finite ϵ-net. If this is not the case then there exists x_2 such that $d(x_2, x_1) \geqslant \epsilon$. Now if either $d(x, x_1) < \epsilon$ or $d(x, x_2) < \epsilon$ for all x then $\{x_1, x_2\}$ is a finite ϵ-net. Otherwise there exists x_3 such that $d(x_3, x_1) \geqslant \epsilon$, $d(x_3, x_2) \geqslant \epsilon$. Continuing, on the assumption that we never get a finite ϵ-net, we have (x_n) such that $d(x_n, x_m) \geqslant \epsilon$ for $m \neq n$. But (x_n) obviously has no convergent subsequence, contrary to sequential compactness. Hence we must get a finite ϵ-net, i.e. X is totally bounded.

We now show that X is compact by contradiction. Suppose X is not compact. Then there exists a cover $\{G_\alpha\}$ of X which has no finite subcover. Since X is totally bounded we have, for each n, a cover of X consisting of a finite number of spheres of radius $1/n$ (a finite $1/n$-net). Thus for each $n = 1, 2, \ldots$ at least one of the spheres of the $1/n$-net, call it S_n, cannot be covered by finitely many of the sets of $\{G_\alpha\}$.

Let x_n be the centre of S_n. Then X sequentially compact implies that (x_n) has a convergent subsequence $x_{n_k} \to x \in X$. Since $\{G_\alpha\}$ covers X we have $x \in G_\alpha$ for some α and so the openness of G_α implies the existence of $S(x, r) \subset G_\alpha$. Now by taking k large enough we can ensure that $S(x_{n_k}, 1/n_k) \subset S(x, r)$, whence the sphere $S_{n_k} \subset G_\alpha$ for some α, which is contrary to the fact that S_{n_k} cannot be covered by finitely many of the sets of $\{G_\alpha\}$. Thus X must be compact.

A final property of compact sets that we examine concerns the behaviour of certain sequences of real functions on them. If we first take an arbitrary set S in a metric space (X, d) and continuous real functions $f_n : S \to R$, then it is easy to prove, exactly as in theorem 4, section 3, chapter 1, that:

If $f_n \to f$ uniformly on S then f is continuous on S.

The converse of this is generally false, e.g. $f_n(x) = nx(1-x)^n$ on $[0, 1]$. We now show that a converse can be obtained for compact topological spaces, provided the convergence is monotonic.

Theorem 22. *Let X be a compact topological space and suppose $f_n \to f$ pointwise on X, where f is lower semicontinuous and each f_n is upper semicontinuous on X. Also, let $f_n \geqslant f_{n+1}$ on X. Then $f_n \to f$ uniformly on X.*

Proof. Take any $\epsilon > 0$ and consider

$$G_n = \{x \in X \mid f_n(x) - f(x) < \epsilon\}.$$

Then G_n is open, because $f_n - f$ is upper semicontinuous. Since f_n decreases we have $G_n \subset G_{n+1}$ for every n. Also, $x \in X$ implies

$$0 \leqslant f_n(x) - f(x) < \epsilon$$

for all sufficiently large n, whence $X = \cup\{G_n \mid n \in N\}$. The compactness of X implies that there exists $p(\epsilon)$ such that

$$X = \bigcup_{n=1}^{p} G_n = G_p \subset G_n$$

for every $n > p$. Hence $x \in X$ implies $x \in G_n$ for every $n > p$, i.e.

$$0 \leqslant f_n(x) - f(x) < \epsilon \quad \text{for} \quad n > p,$$

which means precisely that $f_n \to f$ uniformly on X. ∎

Exercises 5

1. Let X be any set and F a collection of subsets of X. Then F is said to have the finite intersection property if any finite subcollection of F has a nonempty intersection.

   Prove that, if X is a topological space, then X is compact if and only if every collection of closed sets with the finite intersection property has a nonempty intersection.

2. A topological space is called countably compact if every countable open cover has a finite subcover. Prove that the continuous image of a countably compact space is countably compact.

3. Let $f: X \to R$ be upper semicontinuous on the countably compact topological space X (see 2, above). Prove that f is bounded above and attains its bound.

4. Let X be a compact metric space and let (x_n) be a sequence in X. If $E_n = \{x_n, x_{n+1}, \ldots\}$ show that $\{\overline{E}_n\}$ is a collection of closed sets with the finite intersection property (see 1, above). Hence show that X is sequentially compact (this gives another proof of theorem 21 (i), section 5, chapter 2).

5. Suppose $f: X \to Y$ is continuous, where X is a compact metric space and Y is a metric space. Prove that f is uniformly continuous on X. (Hint: There exists $\delta = \delta(\epsilon, x) > 0$, by continuity of f. Cover X with half-size spheres $S(x, \delta/2)$ and extract a finite subcover.)

6. In a metric space, prove that a totally bounded set is bounded. By considering $S = \{e_k\}$, $k = 1, 2, \dots$ in l_2, where $e_k = (0, 0, \dots, 1, 0, 0, \dots)$ with 1 in the kth place, show that a set may be bounded but not totally bounded.

6. Category and uniform boundedness

The notion of category, partly through its consequences on uniform boundedness of families of functions, is most important in analysis. For generality we define category for a topological space but for applications we confine ourselves to metric spaces.

First category set

> Let X be a topological space and S a subset of X. Then S is of the first category if and only if it can be expressed as a countable union of nowhere dense sets.

Second category set

> S is of the second category if and only if it is not of the first category, i.e. S cannot be expressed as a countable union of nowhere dense sets.

We note that the term nowhere dense has not been explicitly defined for a topological space, but the definition given in a metric space still applies if we require $A = \sim \bar{S}$ to be dense, i.e. $\bar{A} = X$.

Example 30. The rationals Q are of the first category in the reals R. For Q is the countable union of its elements $q_1, q_2, \dots$, and each set $\{q_i\}$, $i = 1, 2, \dots$, is obviously nowhere dense in R.

An important theorem of Baire tells us that every complete metric space (as a subset of itself) is of the second category. This is theorem 24 below. First we establish theorem 23, from which Baire's theorem readily follows.

Theorem 23. *Let X be a complete metric space and let (G_n) be a sequence of dense open subsets of X. Then $\cap G_n \neq \varnothing$.*

Proof. Take $x_1 \in G_1$. Then there exists a sphere $S(x_1) \subset G_1$. Take a closed sphere $S[x_1, r_1] \subset S(x_1)$. Now G_2 is dense in X and so there exists $x_2 \in G_2 \cap S[x_1, \frac{1}{2}r_1]$. Hence there exists $S(x_2) \subset G_2$ and we then take a closed sphere $S[x_2, r_2]$ whose radius r_2 is less than the radius of $S(x_2)$ and also less than $\frac{1}{2}r_1$. Then $r_2 < \frac{1}{2}r_1$, $S[x_2, r_2] \subset S[x_1, r_1]$ and $S[x_2, r_2] \subset G_2$.

Continuing in this way we find closed spheres $S[x_n, r_n] \subset G_n$ which are nested: $S[x_n, r_n] \subset S[x_{n-1}, r_{n-1}]$ and $r_n \to 0$ $(n \to \infty)$. Since X is complete, we know by theorem 8 that there exists x such that

$$x \in \cap \, S[x_n, r_n] \subset \cap \, G_n.$$

Thus $\cap \, G_n \neq \varnothing$.

Theorem 24 (Baire category theorem). *A complete metric space X is of the second category.*

Proof. Suppose X is not of the second category. Then X is of the first category:

$$X = \overset{\infty}{\underset{n=1}{\cup}} E_n \left(= \overset{\infty}{\underset{n=1}{\cup}} \bar{E}_n \right),$$

where each E_n is nowhere dense. Hence, taking complements,

$$\varnothing = \cap \, (\sim \bar{E}_n),$$

and the sets $G_n = \, \sim \bar{E}_n$ are open (for $\bar{E}_n$ is closed) and dense (since E_n is nowhere dense). By theorem 23 we have that $\cap \, G_n \neq \varnothing$, which gives a contradiction. Thus X is of the second category.

The final theorem of this chapter is usually referred to as the uniform boundedness principle. As we shall later see it has important consequences.

Theorem 25 (Uniform boundedness principle). *Let P be a collection of real lower semicontinuous functions p defined on the second category metric space X and suppose*

$$p(x) \leqslant M(x) < \infty, \quad each\ x \in X, \quad all\ p \in P. \tag{4}$$

Then there exists a sphere S in X and a constant M such that

$$p(x) \leqslant M, \quad each\ x \in S, \quad all\ p \in P. \tag{5}$$

Proof. First we remark that the essential difference between (4) and (5) is that the inequality in (4) involves a bound $M(x)$ which depends on x but not on p and the inequality holds on the whole of X. In (5) the bound M is uniform depending neither on x nor on p, but this time the inequality holds only on a sphere in X, not necessarily on the whole of X. Usually the family P will be a sequence (p_n) of functions.

For the proof we first consider, for each $p \in P$ and each positive integer m, the set

$$E(m, p) = \{x \,|\, p(x) \leqslant m\}.$$

This set is closed, since $\sim E(m, p) = \{x \,|\, p(x) > m\}$ and p is lower semi-continuous (see section 4, chapter 2). It follows that

$$E_m \equiv \cap \{E(m, p) \,|\, p \in P\},$$

being an intersection of closed sets, is closed. Now

$$X = \cup \{E_m \,|\, m = 1, 2, \ldots\}. \tag{6}$$

For if $x \in X$ then $p(x) \leqslant M(x)$ for all $p \in P$ and so there is an integer $m(x)$ such that $p(x) \leqslant m(x)$ for all $p \in P$. This implies that $x \in E_{m(x)}$, which proves (6).

By hypothesis X is of the second category, so that (6) implies that at least one of the sets E_m, say E_M, is not nowhere dense (if all the E_m were nowhere dense then the countable union of them would be of the first category). Since E_M is not nowhere dense we have that $\bar{E}_M$ contains some sphere S and the fact that E_M is closed implies that $S \subset E_M = \bar{E}_M$.

Finally, $x \in S$ implies $x \in E_M$, which implies $p(x) \leqslant M$ for all $p \in P$ and this proves (5).

Corollary 1. *The conclusion holds if we replace P by a collection F of continuous real functions f.*

Corollary 2. *The result of the theorem holds if X second category is replaced by X complete.*

Proof. X complete implies X second category, by theorem 24.

In chapter 4 we shall give a far-reaching application of theorem 25, which is known as the Banach–Steinhaus theorem for normed spaces. This theorem has many uses in analysis, some of which are rather unexpected. For example, we shall use it to prove the existence of a continuous periodic function whose Fourier series diverges at a point.

Exercises 6

1. Show that any subset of a first category set is a first category set.

2. Prove that $\bigcup\limits_{n=1}^{\infty} E_n$ is of the first category when each E_n is first category.

3. Prove that the set Z of integers is of the first category in R but that Z is of the second category in itself.

4. Let $f_n : R \to R$ be continuous and suppose $f_n(x) \to f(x)$ $(n \to \infty)$ pointwise on R. Prove that there exists a set S of the first category in R such that f is continuous on $\sim S$. As a hint, define

$$F_{m,n} = \{x \in R | \, |f_n(x) - f_k(x)| \leqslant m^{-1}, \text{ for all } k \geqslant n\}.$$

Then consider $$S = \bigcup_{n,m} (F_{m,n} \sim F_{m,n}^{\circ}),$$

where $F_{m,n}^{\circ}$ is the interior of $F_{m,n}$.

3

LINEAR AND LINEAR METRIC SPACES

1. Linear spaces

Several spaces, of interest in both analysis and in algebra, were given in chapter 2. There it was emphasized that any algebraic structure that a space had was not relevant as long as one was concerned with purely metric properties. Of course it would be foolish not to fully exploit any natural structure that a space might have. In most of the examples of chapter 2 it is possible to define addition of elements of the space and also multiplication of elements of the space by real or complex numbers. It is usual to call elements of the space *vectors*, and real or complex numbers *scalars*. As an example we may take the space s, of all sequences $x = (x_n)$ of complex numbers. The natural way of adding sequences is co-ordinatewise:

$$x + y = (x_n) + (y_n) = (x_n + y_n).$$

Equally naturally we multiply a sequence x by a complex scalar λ: $\lambda . x = (\lambda x_n)$. Taking these as our definitions of $+$, which is an operation on $s \times s$, and $.$, which is an operation on $C \times s$, we obviously have such properties as $x + y = y + x$, $\lambda(x + y) = \lambda x + \lambda y$, $\lambda(\mu x) = (\lambda \mu)x$, valid for all x, y in s and all λ, μ in C. As is usual we omit the operation $.$ in $\lambda . x$, merely writing λx. The properties just mentioned all stem, of course, from the properties of C. For example, by our definition, $\lambda x + \lambda y = (\lambda x_n + \lambda y_n)$, and since $\lambda x_n + \lambda y_n = \lambda\{x_n + y_n\}$ for complex λ, x_n, y_n, we have $(\lambda\{x_n + y_n\}) = \lambda(x_n + y_n) = \lambda(x + y)$, again by our definition of addition and scalar multiplication in s. With the operations defined the space s becomes a *complex linear space*.

Now we wish to define the concept of a general abstract linear space over a scalar field. Usually our scalars will be complex numbers. Occasionally they may be real numbers, but unless specific mention is made to the contrary the scalar field will always be C.

Linear space

*A linear space over C (complex linear space) is a nonempty set X
with a function + on X × X into X, and a function . on C × X
into X such that for all complex λ, μ and elements (vectors)
x, y, z in X we have* (1) $x + y = y + x$, (2) $(x + y) + z = x + (y + z)$,
(3) *there exists* $\theta \in X$ *such that* $x + \theta = x$, (4) *there exists* $-x \in X$
such that $x + (-x) = \theta$, (5) $1.x = x$, (6) $\lambda(x + y) = \lambda x + \lambda y$,
(7) $(\lambda + \mu)x = \lambda x + \mu x$, (8) $\lambda(\mu x) = (\lambda \mu)x$.

An equivalent way of defining a linear space is to say that it is an
additive Abelian group, i.e. + is defined and (1)–(4) hold, for which
also scalar multiplication is defined such that (5)–(8) hold. The element
θ is variously called the zero, neutral element or the origin in X. It is
easy to see that θ and $-x$ are unique. For example, if θ' is also a zero,
then $\theta + \theta' = \theta$. But if θ is a zero, then $\theta + \theta' = \theta'$, whence $\theta = \theta'$. It is
clear that a linear space is quite a rich algebraic structure, being an
Abelian (i.e. commutative) group with respect to the internal operation of +, and also having other properties relating the internal + and
the external scalar multiplication .. For most of the situations we
shall be concerned with it is usual to think of X as the primary object
of interest, together with +. The scalar field C, although an essential
part of the definition of a linear space, may be thought to lurk in the
background.

In future, when we speak of a linear space we shall mean a complex
linear space. If the scalars $\lambda, \mu, \ldots$ are to be real then we shall explicitly
use the term real linear space. The term rational linear space will also
be used, where of course the $\lambda, \mu, \ldots$ will be restricted to be in Q.

Before we continue we should remark that the whole of sections
1 and 2 of this chapter are purely algebraic. It is not until section 3 that
we shall combine a linear space structure with a metric structure in a
special way to obtain a linear metric space. This object, with its
generalization to a topological linear space, or its specialization to a
normed linear space, provides one of the most interesting and fruitful
studies in the whole of mathematics.

Now we give some illustrations of linear spaces.

Example 1. (i) C is a complex linear space with the usual addition and
multiplication for complex numbers. Of course C is much more than
a linear space, it is a field. It is clear that any field is a linear space over
itself.

(ii) R, with the usual addition and multiplication, is a real linear space and a rational linear space, but not a complex linear space.

(iii) R^n becomes a real linear space if we define co-ordinatewise operations as follows: $x+y = (x_1+y_1, ..., x_n+y_n)$, $\lambda x = (\lambda x_1, ..., \lambda x_n)$, where $x = (x_1, ..., x_n)$, $y = (y_1, ..., y_n)$ and λ is real. The axioms of our definition of real linear space are readily checked: $\theta = (0, 0, ..., 0)$ and $-x = (-x_1, -x_2, ..., -x_n)$.

Similarly, C^n may be turned into a complex linear space.

(iv) s becomes linear under the definitions $(x_n)+(y_n) = (x_n+y_n)$, $\lambda(x_n) = (\lambda x_n)$. When we regard any sequence spaces as linear spaces we shall always take the linear operations to be defined as they are in s.

(v) The sequence spaces l_p, c_0, c, l_∞ are all linear spaces with the co-ordinatewise operations of (iv) above. All we have to check really is closure under $+$ and $.$, i.e. $\lambda x + \mu y$ is in the space whenever x, y are in the space. In c, for example, this follows from the elementary theorem in analysis that $x_n \to l$, $y_n \to m$ implies $\lambda x_n + \mu y_n \to \lambda l + \mu m$, as $n \to \infty$, i.e. $(\lambda x_n + \mu y_n) = \lambda x + \mu y$ is in c, for any complex λ, μ.

As another example, consider l_p $(p > 1)$. That $\lambda x \in l_p$ is trivial: $\Sigma |\lambda x_p|^p = |\lambda|^p \Sigma |x_k|^p < \infty$ for any λ and any $x \in l_p$. Also, by the trivial inequality $|x_k+y_k|^p \leqslant 2^p(|x_k|^p + |y_k|^p)$ we see that $x+y \in l_p$ when $x, y \in l_p$.

(vi) Let X be any nonempty set and denote by S the set of all complex functions $f : X \to C$. Define $\lambda f_1 + \mu f_2$ by

$$(\lambda f_1 + \mu f_2)(x) = \lambda_1 f_1(x) + \mu f_2(x),$$

for each $x \in X$. This is the usual co-ordinatewise definition of addition and scalar multiplication for complex functions. It is easy to see that S is a linear space.

(vii) $C[0, 1]$ of example 7, chapter 2, is a linear space. Here the metric structure is not involved. The linear operations are defined as in (vi) above, and $\lambda f_1 + \mu f_2 \in C[0, 1]$ whenever $f_1, f_2 \in C[0, 1]$ is a simple. result on continuous functions from elementary analysis.

(viii) The spaces A and I of example 9, chapter 2, are both linear. For, if f, g are analytic then so is $\lambda f + \mu g$.

These examples show that a wide variety of interesting sets may be made into linear spaces in a perfectly natural manner.

We close this introductory section with the definition of isomorphic linear spaces.

Isomorphism

An isomorphism f between linear spaces (over the same scalar field) is a bijective linear map, i.e. f is bijective and

$$f(\lambda x_1 + \mu x_2) = \lambda f(x_1) + \mu f(x_2).$$

Two linear spaces are called isomorphic (or linearly isomorphic) if and only if there exists an isomorphism between them.

We regard isomorphic linear spaces as equivalent from the algebraic linear space point of view, for an isomorphism clearly preserves the linear operations.

Example 2. Let c be the linear space of convergent sequences and γ the space of convergent series (see exercises 1, qn. 9, chapter 2). With the usual linear operations in sequence spaces, γ is linearly isomorphic to c. To see this we define $f : \gamma \to c$ by $f(a) = A$, where $a = (a_k) \in \gamma$ and $A = (A_k)$, with $A_k = a_1 + a_2 + \dots + a_k$, the kth partial sum of Σa_k. The linearity of f is a consequence of the fact that

$$\sum_{k=1}^{n} \{\lambda a_k + \mu b_k\} = \lambda A_n + \mu B_n$$

for every n. Now if $f(a) = f(b)$ then $A_k = B_k$ for every k and so $a_1 = b_1$, $a_2 = b_2, \dots$, i.e. $a = b$. Thus f is injective. Finally, we must show that f is surjective. If $A \in c$ is given we take $a_1 = A_1$, $a_k = A_k - A_{k-1}$ for $k > 1$. Then

$$\sum_{i=1}^{k} a_i = A_k \to l \quad (k \to \infty)$$

and so Σa_i converges, i.e. $a \in \gamma$. Clearly $f(a) = A$, whence f is surjective.

Other examples of isomorphic linear spaces will occur in later sections.

Exercises 1

1. Use the axioms for a linear space to show that $-x$ is unique; $\lambda\theta = \theta$; $0 . x = \theta$; $(-1)x = -x$; and $\lambda x = \theta$ implies $\lambda = 0$ or $x = \theta$.
2. Find which of the following are linear spaces: (i) the set of all sequences $x = (x_k)$ such that $x_k \to 1$ $(k \to \infty)$, (ii) the set of all polynomials of the form $a_0 + a_1 z + \dots + a_n z^n$, for arbitrary degree n, (iii) the set of all polynomials of degree less than 5, (iv) the set of all analytic functions $f = f(z)$ satisfying the differential equation $f''(z) - f'(z) - 2 = 0$.
3. Show that the set of all numbers of the form $a + b\sqrt{2}$; $a, b \in Q$, is a rational linear space. Is it a real linear space?
4. Prove that $f : R^3 \to R^3$, given by $f(x) = (x_2, -x_1, x_3)$ is an isomorphism. It is usual to call an isomorphism of a linear space into itself an automorphism.
5. Attempt to show that R^1 and R^2 are not isomorphic.

6. Let X be a topological space and let $f: X \to C$. Then the support of f is defined as the closure of the set $\{x \in X | f(x) \neq 0\}$ and is written supp (f). Thus supp (f) is a closed set. Denote by S the set of all continuous functions $f: X \to C$ whose support is compact. Prove that S is a linear space. As a hint, first show that supp $(f+g) \subset$ supp $(f) \cup$ supp (g).

7. Prove that $l(p) = \{x = (x_k) | \Sigma |x_k|^{p_k} < \infty\}$, where $p_k > 0$, is a linear space if and only if sup $p_k < \infty$.

8. Let E be a subset of s and let

$$E' = \{y \in s | \Sigma x_k y_k \text{ converges for all } x \in E\}.$$

Prove that E' is a linear space. Find E' when (i) $E = \{\theta\}$, (ii) $E = s$, (iii) $E = l_\infty$.

2. Subspaces, dimensionality, factorspaces, convex sets

In this section we introduce several simple ideas which will be useful in the sequel. The main aims are to show that C^n is essentially the only n-dimensional linear space (i.e. every n-dimensional linear space is isomorphic to C^n), and also that every linear space has a Hamel base. Some preliminary definitions will now be made. Throughout, X will denote a linear space.

Subspace

A subset M, or linear manifold, in a linear space X is a nonempty subset of X such that $\lambda x + \mu y \in M$ whenever $x, y \in M$, for all $\lambda, \mu \in C$.

We note that if $\{M_\alpha\}$ is a family of subspaces then $\cap M_\alpha$ is also a subspace.

Linear hull

Let S be a subset of the linear space X. Then l. hull (S), the linear hull of S, is the intersection of all subspaces containing S.

We remark that other authors use such terms as 'span of S' or 'subspace generated by S' for what we call l. hull (S).

$A + B$, λA

Let A, B be subsets of X. Then we define
$$A + B = \{x + y | x \in A, y \in B\},$$
$$\lambda A = \{\lambda x | x \in A\}.$$
It is usual to write $A + y$ instead of the more precise $A + \{y\}$.

Linear independence

A finite subset $\{x_1, ..., x_n\}$ of X is called a linearly independent set if and only if a relation of the form

$$\lambda_1 x_1 + ... + \lambda_n x_n = \theta$$

implies $\lambda_1 = \lambda_2 = ... = \lambda_n = 0$. An expression of the form $\lambda_1 x_1 + ... + \lambda_n x_n$ is called a linear combination of the vectors $x_1, ..., x_n$.

By special convention we shall regard the empty set $\varnothing$ as linearly independent. If a finite subset is not linearly independent it will be called linearly dependent.

An arbitrary subset (not necessarily finite) of X is called linearly independent if and only if every one of its finite subsets is linearly independent.

The following idea of a base, i.e. a linearly independent set which generates a space, is due to the German mathematician Hamel.

Hamel base

A subset B of X is called a Hamel base for X if and only if B is a linearly independent set and $1.\mathrm{hull}\,(B) = X$.

Dimensionality

A linear space X is called finite dimensional if and only if X has a finite Hamel base B, i.e. B is a finite set which is a Hamel base. If X is not finite dimensional it is called infinite dimensional.

Later in this section we shall show that every linear space has a Hamel base. In the case when the space is finite dimensional, we show in theorem 2 that all Hamel bases have the same number of elements. This enables us to define the dimension of a finite dimensional space as the number of elements in any Hamel base. In the infinite dimensional case it can be shown that all Hamel bases are equivalent as sets (see chapter 1, section 1). One may then define the dimension of an infinite dimensional space as the cardinal of its Hamel base. However, we do not wish to pursue this matter and we shall content ourselves with a discussion of finite dimensionality.

Now we give some examples connected with our definitions.

Example 3. (i) l_p $(p \geqslant 1)$, c_0, c, l_∞ are all subspaces of s.

(ii) The set of all polynomials on $[0, 1]$ is a subspace of $C[0, 1]$.

(iii) $1.\mathrm{hull}(\varnothing) = \{\theta\}$. For the intersection of all subspaces containing $\varnothing$ is precisely the intersection of all subspaces, but θ belongs to every subspace and also $\{\theta\}$ is itself a subspace.

If $S \neq \varnothing$ it is easy to show that $1.\mathrm{hull}(S) =$ the set of all finite linear combinations of elements of S.

(iv) For any set A we have $2A \subset A + A$. For $x \in 2A$ implies $x = 2a = (1+1)a = 1.a + 1.a = a + a$, where $a \in A$. Note we have used axioms (7) and (5) of the definition of linear space. Obviously the inclusion $2A \subset A + A$ can be strict.

(v) Consider the linear space C^n and let $e_i = (0, 0, ..., 1, 0, ...)$, where 1 is in the ith place and there are zeros in the other $n-1$ places. Then $\{e_1, e_2, ..., e_n\}$ is called the set of unit vectors in C^n. This set is linearly independent, since

$$\lambda_1 e_1 + ... + \lambda_n e_n = (\lambda_1, ..., \lambda_n)$$

and so

$$\lambda_1 e_1 + ... + \lambda_n e_n = \theta$$

is equivalent to

$$(\lambda_1, ..., \lambda_n) = (0, ..., 0),$$

whence

$$\lambda_1 = \lambda_2 = ... = \lambda_n = 0.$$

(vi) Let $X = \{\theta\}$ be a trivial linear space, consisting of the single element θ. Then the empty set $\varnothing$ is a Hamel base for X—by convention $\varnothing$ is linearly independent, and $1.\mathrm{hull}(\varnothing) = \{\theta\}$ from (iii) above.

(vii) B, the set of unit vectors in C^n, is a Hamel base for C^n. We know already that B is linearly independent and if $x = (x_1, ..., x_n) \in C^n$, then $x = x_1 e_1 + ... + x_n e_n$, a linear combination of elements of B. Hence $x \in 1.\mathrm{hull}(B)$, so that $C^n = 1.\mathrm{hull}(B)$. Since B is a finite set it follows that C^n is finite dimensional. Observe that B is not the only Hamel base for C^n. For example, in C^2, the set $\{(1, 1), (0, -1)\}$ is a Hamel base, as is readily checked.

So far our classification of linear spaces has been into finite dimensional and infinite dimensional. To actually define the *dimension* of a finite dimensional space we need the following two theorems.

Theorem 1. *Let X have a Hamel base with n elements. Then any set of $n+1$ elements in X is linearly dependent.*

Proof. If $n = 1$ and $\{b\}$ is a Hamel base, then for each x_1, x_2 in X we have $x_1 = \lambda_1 b$, $x_2 = \lambda_2 b$. If $\lambda_1 \lambda_2 = 0$ then either $x_1 = \theta$ or $x_2 = \theta$, so

$\{x_1, x_2\}$ is linearly dependent. If $\lambda_1 \lambda_2 \neq 0$ then $\lambda_1 \neq 0$, $\lambda_2 \neq 0$ and again $\{x_1, x_2\}$ is linearly dependent: $\lambda_2 x_1 - \lambda_1 x_2 = \lambda_2 \lambda_1 b - \lambda_1 \lambda_2 b = \theta$. This deals with the case $n = 1$. Now we consider the case $n = 2$, finishing the proof by induction.

Take $n = 2$, $B = \{b_1, b_2\}$ a Hamel base. Let $S = \{x_1, x_2, x_3\}$ be any 3 element set in X. Then

$$x_i = \lambda_{i1} b_1 + \lambda_{i2} b_2 \quad (i = 1, 2, 3).$$

Now consider the subspace $M = 1. \text{hull} \{b_1\}$. If all of $x_1, x_2, x_3 \in M$, then, since $\{b_1\}$ is a Hamel base for M, the case $n = 1$ shows that the 2 element set $\{x_2, x_3\}$ is linearly dependent; *a fortiori* S is linearly dependent. If however x_1, x_2, x_3 are not all in M we may relabel so as to ensure that x_3 is not in M. This implies $\lambda_{32} \neq 0$, for otherwise $x_3 = \lambda_{31} b_1 \in M$, contrary to hypothesis. Now for $i = 1, 2$,

$$y_i = x_i - \frac{\lambda_{i2}}{\lambda_{32}} x_3 = \lambda_{i1} b_1 + \lambda_{i2} b_2 - \frac{\lambda_{i2}}{\lambda_{32}} (\lambda_{31} b_1 + \lambda_{32} b_2)$$

which belongs to M. From the case $n = 1$, the two element set $\{y_1, y_2\}$ is linearly dependent, i.e. there exist μ_1, μ_2 not both zero such that $\mu_1 x_1 + \mu_2 x_2 + \lambda x_3 = \theta$, where λ depends on μ_1, μ_2, λ_{12}, λ_{22}, λ_{32}, its precise form being irrelevant. Hence we see that S is linearly dependent which proves the theorem for the case $n = 2$.

Using the idea of the case $n = 2$ it is easy to finish the proof inductively. This is left as an exercise.

Theorem 2. *Let X be finite dimensional. Then all the Hamel bases for X have the same number of elements.*

Proof. Suppose B is a Hamel base, with n elements. Let B' be another Hamel base for X. B' must be finite, otherwise it would have $n+1$ linearly independent elements, contrary to theorem 1. If B' has m elements, say, then we must have $m = n$. For if $m > n$ or $m < n$ we contradict theorem 1, since B, B' are both bases.

In view of theorem 2 the following definition is meaningful.

Dimension

If X is a finite dimensional space then its dimension is defined to be the number of elements in any of its Hamel bases.

Example 4. C^n has dimension n, since $\{e_1, ..., e_n\}$ is a Hamel base with n elements.

The next theorem shows that example 4 is in a sense the only possible example that can be given of an n-dimensional linear space.

Theorem 3. *If X is finite dimensional, with dimension n, then X is isomorphic to C^n.*

Proof. Since X is finite dimensional there is a Hamel base $\{b_1, ..., b_n\}$. If $x \in X$ then $x = \lambda_1 b_1 + ... + \lambda_n b_n$ for some scalars λ_i. The λ_i are unique, for if $x = \mu_1 b_1 + ... + \mu_n b_n$ then

$$(\lambda_1 - \mu_1)b_1 + ... + (\lambda_n - \mu_n)b_n = \theta,$$

whence $\lambda_i = \mu_i \, (1 \leqslant i \leqslant n)$, by the linear independence of the b_i. It follows now that the map $f : X \to C^n$, given by

$$f(x) = (\lambda_1, ..., \lambda_n)$$

is well-defined. It is clearly bijective and also it is easy to check that $f(\alpha x + \beta y) = \alpha f(x) + \beta f(y)$ for scalar α, β and $x, y \in X$. Hence f is an isomorphism.

For examples of infinite dimensional spaces we may cite the sequence spaces c_0, s, and the function space $C[0, 1]$. For example, if e_k denotes the infinite sequence with 1 in the kth place and 0 elsewhere, then $\{e_1, e_2, ...\}$ is an infinite linearly independent set in c_0, so whatever n we take there are always $n + 1$ linearly independent elements in c_0.

By definition, a finite dimensional space has a finite Hamel base. It is by no means obvious that an infinite dimensional space has a Hamel base at all. To show the existence of Hamel bases for all linear spaces it seems to be necessary to employ Zorn's lemma, or some form of the axiom of choice. Regarding, as we do, Zorn's lemma as an axiom it is quite easy to show that every linear space has a Hamel base. However, the process employed is highly non-constructive and provides no explicit Hamel base for any specific space.

Theorem 4. *Every linear space X has a Hamel base.*

Proof. Let P be the class of all linearly independent subsets of X. Then P is nonempty since $\varnothing \in P$. Partially order P by set inclusion and let $T = \{L_\alpha\}$ be a totally ordered subset of P. Let $L = \cup L_\alpha$ be the union of all the L_α in T. Then L is a linearly independent set. For if $S = \{s_1, ..., s_n\}$ is a finite subset of L then $s_i \in L_{\alpha_i} \, (1 \leqslant i \leqslant n)$. By the total order in T we may arrange, by relabelling if necessary, that $L_{\alpha_1} \subset ... \subset L_{\alpha_n}$. Hence $s_i \in L_{\alpha_n} \, (1 \leqslant i \leqslant n)$, so that S is linearly independent, being a finite subset of the linearly independent L_{α_n}. Thus T has L for an upper bound. By Zorn's lemma (see chapter 1), P has a maximal element, B, say. Now every x in X is also in l. hull (B). For if not then there exists an $x \in \sim$ l. hull (B), whence $B \cup \{x\}$ is linearly independent. But $B \cup \{x\} \supset B$, strictly, contrary to the fact that B is maximal. Hence we have shown that B is linearly independent and that l. hull $(B) = X$, whence B is the required Hamel base.

The reader who is familiar with the elementary theory of groups will have encountered the idea of a factorgroup or quotient group G/H, where G is a group and H is a normal subgroup. If X is a linear space and M is a subspace one may define a factorspace X/M, which is somewhat analogous to a factorgroup G/H. We shall show that natural operations can be defined in X/M so as to make it into a linear space. Thus, each subspace M of X generates a new linear space X/M.

Factorspace X/M

Let X be a linear space and M a subspace. Define $x \equiv y \pmod{M}$ to mean that $x - y \in M$. Then $\equiv$ is an equivalence relation and X/M, the factorspace X modulo M, is defined to be the set $\{E_x | x \in X\}$, where $E_x = \{y | y \equiv x \pmod{M}\}$. With the definitions $E_x + E_y = E_{x+y}$, $\lambda E_x = E_{\lambda x}$, X/M becomes a linear space.

The verification of our assertions is routine. That $\equiv$ is an equivalence relation is trivial. Also, our definitions of addition and scalar multiplication in X/M are meaningful. For example, if $x \equiv x' \pmod{M}$, $y \equiv y' \pmod{M}$ then $x + y \equiv x' + y' \pmod{M}$, so that picking different representatives from E_x, E_y does not alter $E_x + E_y$. Similarly one sees that λE_x is well defined. Since X is linear it is clear that X/M satisfies the axioms for a linear space. We observe that the zero in X/M is the subspace M. For $E_\theta = \{y \equiv \theta \pmod{M}\} = \{y \in M\} = M$.

Example 5. Let $X = L[0, 1]$ (see example 6, chapter 2). With the usual algebraic operations for combining functions X is a linear space. Write $M = \{f \in X | f(t) = 0 \text{ almost everywhere}\}$. Then M is a subspace of X, and $f \equiv g \pmod{M}$ means $f(t) = g(t)$ almost everywhere. In integration theory it is usual to work with the factorspace X/M, rather than with X itself. Such statements as $f = g$ are usually interpreted, in integration theory, as $f \equiv g \pmod{M}$.

The concepts of convex and absolutely convex set in a linear space are important in several connections. These we now define.

Convexity and absolute convexity

Let E be a subset of a linear space. Then

(i) *E is called convex if and only if $x, y \in E, \lambda + \mu = 1, \lambda \geqslant 0$, $\mu \geqslant 0$ imply $\lambda x + \mu y \in E$.*

(ii) *E is called balanced if and only if $x \in E$, $|\lambda| \leqslant 1$ imply $\lambda x \in E$.*

(iii) *E is called absolutely convex if and only if $x, y \in E$, $|\lambda| + |\mu| \leqslant 1$ imply $\lambda x + \mu y \in E$.*

It is not hard to show that a set E is absolutely convex if and only if it is convex and balanced. The proof of this is left as an exercise.

There is an obvious geometrical way of describing convexity. Let us call $L(x, y) = \{\lambda x + (1 - \lambda)y \,|\, 0 \leqslant \lambda \leqslant 1\}$ the line segment joining x and y. Then it is clear that E is convex if and only if E contains $L(x, y)$ whenever it contains x and y.

Example 6. (i) Trivially, every subspace is absolutely convex and also every absolutely convex set is convex.

(ii) Let d be the usual metric on C^n. Then every sphere $S(a, r)$ in C^n is convex. For if $d(x, a) < r$, $d(y, a) < r$, $\lambda + \mu = 1$, etc., then, by Minkowski's inequality,

$$d(\lambda x + \mu y, a) = \left(\sum_{1}^{n} |\lambda(x_k - a_k) + \mu(y_k - a_k)|^2 \right)^{\frac{1}{2}}$$

$$\leqslant \lambda d(x, a) + \mu d(y, a) < r.$$

There is a simple way of generating absolutely convex sets using certain types of real functions called seminorms.

Seminorms

A seminorm p, on a linear space X, is a function $p : X \to R$ such that (1) $p(\lambda x) = |\lambda| p(x)$, (2) $p(x + y) \leqslant p(x) + p(y)$.

Property (1) is called absolute homogeneity of p, and property (2) is called subadditivity of p. Thus a seminorm is a real, subadditive, absolutely homogeneous function on X. Moreover, by (1) and (2), $0 = p(\theta) \leqslant p(x) + p(-x) = 2p(x)$, whence p is always non-negative.

Example 7. (i) $p(x) = |x|$ is a seminorm on C.

(ii) $p_1(x) = \sup |x_n|$ and $p_2(x) = |\lim x_n|$ are seminorms on c.

(iii) If $f: X \to C$ is a linear map, then $p(x) = |f(x)|$ is a seminorm on X.

We now have a very simple theorem.

Theorem 5. *Let p be a seminorm on a linear space X. Let $r > 0$. Then the sets $\{x | p(x) < r\}$, $\{x | p(x) \leqslant r\}$ are absolutely convex.*

Proof. Suppose $p(x) \leqslant r$, $p(y) \leqslant r$. Then

$$p(\lambda x + \mu y) \leqslant |\lambda| p(x) + |\mu| p(y) \leqslant (|\lambda| + |\mu|) r \leqslant r,$$

whenever $|\lambda| + |\mu| \leqslant 1$. Hence $\{x | p(x) \leqslant r\}$ is absolutely convex; the other set may be dealt with similarly.

In the next section, when we have introduced the idea of a topological linear space, we shall prove a kind of converse to theorem 5. Starting with a nonempty absolutely convex open set A one can generate a seminorm on X, called the gauge of A (see theorem 6).

Exercises 2

1. Prove that the intersection of any collection of subspaces is a subspace.
2. Let S be a nonempty subset of a linear space. Prove that l. hull (S) is the set of all finite linear combinations of elements of S. Show also that l. hull (S) is the smallest subspace containing S.
3. Show that $S \neq \varnothing$ is a subspace if and only if $S + S \subset S$ and $\lambda S \subset S$ for each λ.
4. Let M_1, M_2 be subspaces. Prove that l. hull $(M_1 \cup M_2) = M_1 + M_2$.
5. Show that the inclusion $2A \subset A + A$ can be strict.
6. Let Φ denote the set of finite sequences $x = (x_n)$. (A sequence x is called finite if and only if there exists $p \in N$ such that $x_n = 0$ for all $n \geqslant p$.) Show that Φ is a subspace of s and that l. hull $\{e_1, e_2, \ldots\} = \Phi$. Prove that the countably infinite set $\{e_1, e_2, \ldots\}$ is a linearly independent subset of s but is not a Hamel base for s.
7. Let M be a subspace of X. Define $f: X \to X/M$ by $f(x) = E_x$. Prove that f, the so-called canonical mapping of X onto X/M, is linear and surjective.
8. Prove that a set is absolutely convex if and only if it is convex and balanced.
9. Let E be convex and A, B absolutely convex. Prove that $x + \lambda E$ is convex and that $A + B$, λA are absolutely convex.
10. Show that any intersection of convex sets is convex. Show also that the smallest convex set containing a given set S is the set of all finite linear

combinations $\Sigma\lambda_i s_i$ with $\lambda_i \geqslant 0$, $\Sigma\lambda_i = 1$, where $s_i \in S$. This smallest convex set is called the convex hull of S.

11. Let A be a nonempty absolutely convex set. Prove that $\theta \in A$ and that $|\lambda| \leqslant |\mu|$ implies $\lambda A \subset \mu A$.

3. Linear metric spaces, paranorms, seminorms and norms

We have now reached the stage when it is natural to combine the metric ideas of chapter 2 with the linear space concepts of the present chapter. No significant advance can be made however unless the metric and linear properties are linked together in a fairly intimate way. The accepted way of making this link is through continuity. Because of the tremendous number of concrete situations that this linking process covers it is generally recognized to be the most satisfactory method of fusing metric and linear structures.

As far as making definitions is concerned it is no harder to define a linear topological space than it is to define a linear metric space, so this we shall do. However, in this introductory work, we shall not be able to pursue the theory of linear topological spaces—the interested reader is advised to consult the Cambridge Tract by Robertson and Robertson (1964). Before we make a general definition we may note an example which is essentially the source of it. The space C is a metric space with $d(x, y) = |x - y|$; $x, y \in C$, and also C is a linear space over itself, with the usual addition and multiplication. The metric and linear structures are linked by continuity, in that $x \to x_0$, $y \to y_0$, $\lambda \to \lambda_0$ imply $x + y \to x_0 + y_0$ and $\lambda x \to \lambda_0 x_0$, i.e. addition and scalar multiplication are continuous operations. All we require of our linear topological space is that it should enjoy this same property as C. Hence we define a

Linear topological space

A linear topological space is a linear space X which has a topology T, such that the algebraic operations of addition and scalar multiplication in X are continuous. If T is given by a metric we speak of a linear metric space.

Continuity of addition means that $f: X \times X \to X$, defined by $f(x, y) = x + y$ is continuous on $X \times X$, and continuity of scalar multiplication means that $f: C \times X \to X$, defined by $f(\lambda, x) = \lambda x$ is continuous on $C \times X$. The topologies on the products $X \times X$ and $C \times X$

are defined as follows. For any two general topological spaces X_1 and X_2 we say that a set G is open in $X_1 \times X_2$ if and only if, for any $g \in G$, there exists G_1 open in X_1 and G_2 open in X_2 such that $g \in G_1 \times G_2 \subset G$. It is easy to check that the collection of such sets G is in fact a topology for $X_1 \times X_2$. This topology is called the product topology. In our definition of a linear topological space we understand that C has its usual modulus metric topology.

If (X_1, d_1) and (X_2, d_2) are metric spaces then

$$d((x_1, x_2), (y_1, y_2)) = d_1(x_1, y_1) + d_2(x_2, y_2),$$

with (x_1, x_2) and (y_1, y_2) in $X_1 \times X_2$, is a metric on $X_1 \times X_2$. It is a simple matter to show that the metric topology generated by d is exactly the product topology on $X_1 \times X_2$. Thus, if (X, d) is a linear metric space, then, for example, the continuity of λx can be expressed as follows. For each (λ_0, x_0) and for any $\epsilon > 0$, there exists $\delta > 0$ such that $d(x, x_0) + |\lambda - \lambda_0| < \delta$ implies $d(\lambda x, \lambda_0 x_0) < \epsilon$.

Example 8. (i) Let $0 < p_k \leqslant 1$ and let $X = l(p)$. We know already that X is a linear space and also that

$$d(x, y) = \Sigma |x_k - y_k|^{p_k}$$

defines a metric topology on X. Let us show that X is a topological linear space. Continuity of addition is easy. For, omitting the subscript k in the summations, so as to simplify notation, we have

$$d(x + y, a + b) = \Sigma |x + y - (a + b)|^p$$
$$\leqslant \Sigma |x - a|^p + \Sigma |y - b|^p$$
$$= d(x, a) + d(y, b).$$

Continuity of scalar multiplication is handled as follows. Write, again omitting the subscript k,

$$d(\lambda x, \lambda_0 a) = \Sigma |(\lambda - \lambda_0)(x - a) + (\lambda - \lambda_0)a + \lambda_0(x - a)|^p$$
$$\leqslant \Sigma |\lambda - \lambda_0|^p |x - a|^p + \Sigma |\lambda - \lambda_0|^p |a|^p + \Sigma |\lambda_0|^p |x - a|^p$$
$$= \Sigma_1 + \Sigma_2 + \Sigma_3, \quad \text{say}.$$

Now $|\lambda_0|^{p_k} \leqslant \max(1, |\lambda_0|) = M$, say. If $|\lambda - \lambda_0| < 1$ then $|\lambda - \lambda_0|^{p_k} < 1$. Next, let N be a positive integer and let $|\lambda - \lambda_0| < 1$. Then

$$\Sigma_2 \leqslant \sum_{k=1}^{N} |\lambda - \lambda_0|^p |a|^p + \sum_{k=N+1}^{\infty} |a|^p = A + B.$$

Let $\epsilon > 0$ be given and first choose N so large that $B < \epsilon/3$. Then choose $0 < \eta < 1$ such that $|\lambda - \lambda_0| < \eta$ implies that the finite sum A is less than $\epsilon/3$. Write $\delta = \min(\eta, \epsilon/3(1+M))$. Then

$$d(x, a) + |\lambda - \lambda_0| < \delta$$

implies $\quad\quad d(\lambda x, \lambda_0 a) \leqslant \Sigma_1 + A + B + \Sigma_3$

$$\leqslant d(x, a) + \frac{\epsilon}{3} + \frac{\epsilon}{3} + M d(x, a)$$

$$< \epsilon.$$

Thus scalar multiplication is continuous at each point (λ_0, a).

(ii) An example is now given of a space which fails to be a topological linear space only in that continuity of scalar multiplication fails. Let $p = (p_k) = (1/k)$ and write $l_\infty(p) = \{x \mid \sup |x_k|^{p_k} < \infty\}$. Then $l_\infty(p)$ is linear, $d(x, y) = \sup |x_k - y_k|^{p_k}$ is a metric on $l_\infty(p)$ and addition is continuous. However, there exists $x \in l_\infty(p)$ such that $\lambda x \nrightarrow \theta$ as $\lambda \to 0$. For, let $x_k = 1$ for every k. Take $0 < |\lambda| < 1$. Then $|\lambda|^{1/k} < 1$ for all k and $|\lambda|^{1/k} \to 1$ as $k \to \infty$, so that

$$d(\lambda x, \theta) = \sup |\lambda|^{1/k} = 1,$$

whence $\lambda x \nrightarrow \theta$ as $\lambda \to 0$.

We remarked at the end of section 2 of the present chapter that each nonempty open absolutely convex set A generates a seminorm—this we now prove.

Theorem 6. *Let A be a nonempty open absolutely convex set in a topological linear space X. Then*

$$p(x) = \inf\{\lambda > 0 \mid x \in \lambda A\}$$

is a seminorm on X. This seminorm is called the gauge of A.

Proof. $A \neq \varnothing$ implies some $x \in A$ and so $0.x + 0.x \in A$ by absolute convexity, i.e. $\theta \in A$. Since X is a linear topological space we have $\mu x \to \theta$ as $\mu \to 0$, for each $x \in X$. Hence there exists $\delta > 0$ such that $\delta x \in A$, since A is an open set containing θ. Write $\lambda = \delta^{-1}$. Then $x \in \lambda A$, and so for each $x \in X$ there exists $\lambda > 0$ such that $x \in \lambda A$, whence $0 \leqslant p(x) < \infty$.

Now take $x, y \in X$ and let $\epsilon > 0$ be given. Then there exists $\lambda > 0$ such that $x \in \lambda A$ and $\lambda < p(x) + \epsilon$. Also, there exists $\mu > 0$ such that $y \in \mu A$ and $\mu < p(y) + \epsilon$. Since A is absolutely convex we have $x + y \in (\lambda + \mu) A$ and so $p(x + y) \leqslant \lambda + \mu$. Hence

$$p(x+y) < p(x) + p(y) + 2\epsilon,$$

and ϵ is arbitrary, so that $p(x+y) \leqslant p(x)+p(y)$. Finally, one may show that $p(\lambda x) = |\lambda|\, p(x)$ in a similar way, whence p is a seminorm on X. This proves the theorem.

Let us return momentarily to the space $l(p)$ of example 8. This space is a linear metric space. However, it is rather more than this. For the function $g : l(p) \to R$ defined by

$$g(x) = \Sigma\, |x_k|^{p_k}$$

gives the metric d on writing $d(x, y) = g(x-y)$. This function g has the following properties: $g(\theta) = 0$, $g(x) = g(-x)$, g is subadditive and $\lambda x \to \lambda_0 x_0$ whenever $\lambda \to \lambda_0$, $x \to x_0$ (i.e. $g(x-x_0) \to 0$). A function with these properties, defined on a general space is called a

Paranorm

A paranorm $g : X \to R$, X being a linear space, satisfies $g(\theta) = 0$, $g(x) = g(-x)$, $g(x+y) \leqslant g(x)+g(y)$ and $\lambda \to \lambda_0$, $x \to x_0$ imply $\lambda x \to \lambda_0 x_0$. A paranormed space (X, g) is a linear space together with a paranorm g.

It is clear that every paranormed space becomes a linear semimetric space on setting $d(x, y) = g(x-y)$. It can be shown that on every linear semimetric space it is possible to define a paranorm which gives rise to the same semimetric topology. Thus paranormed spaces and linear semimetric spaces are really the same thing. We consider however that the proof of this result is just beyond our modest scope.

We now have two important concrete types of linear topological space: the paranormed space (X, g) and the seminormed space (X, p).

Example 9. (i) $l(p)$, $0 < p_k \leqslant 1$ is a paranormed space with $g(x) = \Sigma\, |x_k|^{p_k}$ (see example 8 (i), p. 83).

(ii) Every seminorm is a paranorm, but not conversely in general. The first statement is easy to verify. For 'not conversely' we consider

$$g(x) = \Sigma\, |x_k|^{1/k}$$

on $l((1/k))$. Then g is a paranorm, but clearly x exists such that $g(2x) < 2g(x)$, so g is not absolutely homogeneous.

An even more specific linear topological space is the normed space. A norm $\|\,.\,\| : X \to R$ is just a seminorm with the additional property $\|x\| = 0$ implies $x = \theta$. Thus a normed space is a pair $(X, \|\,.\,\|)$. Perhaps the simplest example of a normed space is C, with $\|x\| = |x|$.

The theory of normed spaces is one of the basic studies in general analysis and will be developed in chapter 4.

Exercises 3

1. Show that $l(p)$ is paranormed for any bounded $p = (p_k)$. Prove also that $l(p)$ is complete.

   Remark. In general, a complete linear metric, or complete paranormed space, is often called a Fréchet space, after M. Fréchet.

2. Prove that s, with
$$g(x) = \Sigma 2^{-k}(|x_k|/(1+|x_k|))$$
is a Fréchet space.

3. Let (g_n) be a sequence of paranorms on a linear space X. Show that

$$g(x) = \Sigma 2^{-k} g_k(x)/(1 + g_k(x))$$

is a paranorm on X. Prove also that $g(x_n) \to 0$ $(n \to \infty)$ if and only if $g_k(x_n) \to 0$ $(n \to \infty)$ for each k.

4. Let g be a paranorm such that $g(\lambda x) \leqslant |\lambda| \, g(x)$ for all λ and all x. Prove that g is a seminorm.

5. Let p, q be seminorms on X such that $q(x) \leqslant 1$ whenever $p(x) < 1$. Prove that $q(x) \leqslant p(x)$ for all $x \in X$.

6. Let p be the gauge of A. Prove that

$$\{x|p(x) < 1\} \subset A \subset \{x|p(x) \leqslant 1\}.$$

7. Suppose X is finite-dimensional, with Hamel base $\{b_1, \ldots, b_n\}$. Show that

$$\|x\| = \sum_{k=1}^{n} |\lambda_k|$$

is a norm on X, where $x = \lambda_1 b_1 + \ldots + \lambda_n b_n$.

4. Basis

If B is a Hamel base for a linear space X, then B is linearly independent and l. hull $(B) = X$. Hence $x \in X$ implies $x = \Sigma \lambda_k b_k$, a finite linear combination of elements of B. This representation is unique, since B is linearly independent.

In certain problems (see for example section 2, chapter 4) it is useful to have a concept of basis which allows us to express, uniquely, an element x as an infinite series $x = \Sigma \lambda_k b_k$. This idea automatically involves convergence and hence some kind of topology. Of course a Hamel base is free of topology. We shall restrict ourselves to a definition of basis in a paranormed space (X, g) (see section 3 of this chapter).

Basis

Let (X, g) be a paranormed space. A sequence (b_k) of elements of X is called a basis for X if and only if, for each $x \in X$ there exists a unique sequence (λ_k) of scalars such that $x = \sum\limits_{k=1}^{\infty} \lambda_k b_k$, i.e. such that $g\left(x - \sum\limits_{k=1}^{n} \lambda_k b_k\right) \to 0 \ (n \to \infty).$

The idea of a basis was introduced by J. Schauder in 1927; what we call a basis is often termed a Schauder basis. Our choice of terminology makes it clear that we wish to distinguish between basis and Hamel base. However, in finite dimensional spaces these concepts coincide, as is readily proved.

Example 10. Write $e_1 = (1, 0, 0, \dots)$, $e_2 = (0, 1, 0, \dots)$, ... Then (e_k) is a basis for the spaces $l(p)$, c_0 and s, under their natural paranorms or norms:

$$g(x) = (\Sigma |x_k|^{p_k})^{1/M} \quad \text{on } l(p),$$

$$\|x\| = \sup |x_k| \quad \text{on } c_0,$$

$$g(x) = \Sigma 2^{-k}(|x_k|/(1+|x_k|)) \quad \text{on } s.$$

Let us consider $l(p)$, for example. Take any $x = (x_k) \in l(p)$. Write $y_n = x - (x_1, \dots, x_n, 0, 0, \dots)$. Then

$$y_n = x - \sum_{k=1}^{n} x_k e_k = (0, 0, \dots, x_{n+1}, x_{n+2}, \dots).$$

Hence $$[g(y_n)]^M = \sum_{n+1}^{\infty} |x_k|^{p_k} \to 0 \quad (n \to \infty),$$

i.e. $x = \Sigma x_k e_k$. This representation for x is unique. For if

$$x = \Sigma \lambda_k e_k \quad \text{then} \quad g\left(\sum_{1}^{n} (\lambda_k - x_k) e_k\right) \to 0 \quad (n \to \infty),$$

whence $$\sum_{1}^{n} |\lambda_k - x_k|^{p_k} \to 0,$$

which implies $\lambda_k = x_k$ for all k.

We may deal with c_0 and s in the same manner. Note that (e_k) is a basis in c_0 but $\{e_k\}$ is not a Hamel base. For l. hull $\{e_k\} = \Phi$, the space of finite sequences, which is a proper subset of c_0.

Example 11. Take e_k as in example 10 above. Write $e = (1, 1, \ldots)$. Then $(e, e_1, e_2, \ldots)$ is a basis for c, under its natural norm $\|x\| = \sup |x_k|$. For, suppose $x \in c$, with $x_k \to l$ $(k \to \infty)$. Then

$$\left\| x - le - \sum_1^n (x_k - l)\, e_k \right\| = \sup_{k > n} |x_k - l| \to 0 \quad (n \to \infty),$$

so that $x = le + \Sigma(x_k - l)\, e_k$. It is easy to check that this representation for x is unique.

The basis just given for c will be used in section 2, chapter 4, when we determine the dual space of c.

It is known that not all paranormed spaces have a basis. The next result tells us that a paranormed space which has a basis is necessarily a separable semimetric space (see section 3, chapter 2).

Theorem 7. *Let (X, g) be a paranormed space with basis $B = (b_k)$. Then X is separable.*

Proof. Let S be the set of all finite linear combinations $\sum_1^n (r_k + is_k)\, b_k$, with r_k, s_k rational. It is easy to show that S is a countable set. We now show that S is dense in X. Take any $x \in X$. Then $x = \Sigma \lambda_k x_k$, so for each $\epsilon > 0$, there exists N such that

$$g\left(x - \sum_1^N \lambda_k b_k \right) < \epsilon.$$

Take k such that $1 \leqslant k \leqslant N$. Then choose $\mu_k = r_k + is_k$ so close to λ_k that $g((\lambda_k - \mu_k)\, b_k) < \epsilon/N$. This is possible by continuity of scalar multiplication in X. We now have

$$g\left(x - \sum_1^N \mu_k b_k \right) < \epsilon + g\left(\sum_1^N (\lambda_k - \mu_k)\, b_k \right) < 2\epsilon.$$

It follows that S is dense in X, whence X is separable.

Theorem 7 may be used to show that certain paranormed spaces have no basis. For example, it can be shown that l_∞ is not separable, so it has no basis by theorem 7.

Exercises 4

1. Let (X, g) have a basis (b_k). Then $x = \Sigma \lambda_k b_k$, for each $x \in X$. Prove that each map $f_k : X \to C$, given by $f_k(x) = \lambda_k$, is linear on X.
2. Let P denote the normed space of all real polynomials in t on $[0, 1]$ with $\|x\| = \max\{|x(t)| \,|\, 0 \leqslant t \leqslant 1\}$. Prove that $\{t^k | k = 0, 1, \ldots\}$ is a basis for P. Determine the maps f_k of question 1.

3. (i) Prove that (e_k) is a basis for c_0 and show that the sequence

$$\left(\left\| \sum_1^n x_k e_k \right\| \right)$$

increases.

(ii) Let $x \in c$, $\lim x_k = l$. Show that the sequence

$$\left(\left\| le + \sum_1^n (x_k - l) e_k \right\| \right)$$

increases.

4. Prove that l_∞ is normed, with $\|x\| = \sup |x_k|$. Show that $d(x, y) = \|x - y\|$ is a metric on l_∞ and that, in this metric, l_∞ is not separable. Deduce that l_∞ has no basis.

5. Consider $C[0, 1]$ as a normed space of real-valued functions. Let S denote the set of all polygonal functions on $[0, 1]$ whose vertices are at points with rational co-ordinates. Prove that $C[0, 1]$ is separable by showing that S is a countable dense subset.

Attempt to show that $C[0, 1]$ has a basis.

5. Distributions

Distribution theory was initiated, in its modern form, by the Russian mathematician S. L. Soboleff in 1936. It was greatly developed and systematized by L. Schwartz in the 1950s. Distributions, or generalized functions as they are sometimes called, have found application in many fields, e.g. in differential equation theory, eigenfunction expansions of differential operators and also in the theory of random processes.

Here we intend merely to give a few of the underlying ideas of distribution theory so as to prepare the interested reader for the more ambitious and comprehensive works (e.g. Schwartz, 1950; Gelfand and Shilov, 1964; D. S. Jones, 1966).

Test functions

First we define the linear space D of test functions. We say that $x \in D$ if and only if (1) $x : R \to C$, (2) *x is indefinitely differentiable on R and* (3) $x(t) = 0$ *outside some closed interval $[a, b]$. The closed interval $[a, b]$ will generally depend on the function $x = x(t)$.*

Example 12. $x(t) = \exp (t^2 - 1)^{-1}$ for $|t| < 1$ and $x(t) = 0$ for $|t| \geq 1$, defines a test function x. It is easily shown that every derivative $x^{(k)}(t) = 0$ at $t = \pm 1$.

The concept of convergence in D is defined as follows. Let (x_n) be a sequence in D. Then we say that $x_n \to 0$ (in D) if and only if there exists an interval $[a, b]$, the same for all x_n, such that $x_n(t) = 0$ outside $[a, b]$, for all n, and $x_n^{(k)}(t) \to 0$ $(n \to \infty$, uniformly on $[a, b])$ for $k = 0, 1, \ldots$.

Example 13. Take x as in example 12 and let $x_n(t) = x(t)/n$. Then $x_n(t) = 0$ outside $[-1, 1]$, for all n, and $|x_n(t)| \leqslant 1/ne$ for all n and t, whence $x_n(t) \to 0$ (uniformly on R). Now for $k = 1, 2, \ldots, x^{(k)}(t)$ is differentiable on R and so is bounded on $[-1, 1]$. For such k we thus have $|x_n^{(k)}(t)| \leqslant M_k/n$ for all n and t, so that $x_n^{(k)}(t) \to 0$ $(n \to \infty$, uniformly on $[-1, 1])$. Hence $x_n \to 0$ (in D).

Convergence of a sequence (x_n) in D to an element $x \in D$ is defined by $x_n - x \to 0$ (in D). We are now in a position to define the term distribution.

Distribution

A distribution f is a continuous linear functional on D, the linear space of test functions.

To say that f is a functional is to say that $f : D \to C$. Linearity of f means that $f(\lambda x + \mu y) = \lambda f(x) + \mu f(y)$ for any $\lambda, \mu \in C$ and any $x, y \in D$. Finally, f is called continuous if and only if $x_n \to 0$ (in D) implies $f(x_n) \to 0$. We denote the set of all distributions by D'.

Note that, if f is a distribution, then $x_n \to x$ (in D) implies $f(x_n) \to f(x)$.

Example 14 (Delta distribution). Define $\delta(x) = x(0)$ for each $x \in D$. Then it is trivial to check that $\delta \in D'$. The distribution δ is sometimes rather loosely referred to as the 'delta function'.

Example 15. Let F be Lebesgue integrable on every finite interval. Define, on D,

$$f(x) = \int_{-\infty}^{\infty} F(t)\, x(t)\, dt.$$

The integral is really over a finite range since x vanishes outside some closed interval. Thus f is a functional on D, which is linear by the linear properties of the integral. Suppose now that $x_n \to x$ (in D). Then

$$|f(x_n) - f(x)| \leqslant \int_{-\infty}^{\infty} |F(t)|\, |x_n(t) - x(t)|\, dt \to 0 \quad (n \to \infty),$$

since $x_n(t) \to x(t)$ (uniformly on some interval $[a, b]$). We have now shown that each F, which is suitably integrable, generates a distribution f by means of an integral. Any distribution which is generated by some function integrable on every finite interval, by means of an integral, is called a *regular distribution*. Non-regular distributions are called *singular*.

Example 16. The delta distribution of example 14 is singular. For suppose there exists F, Lebesgue integrable on every finite interval, such that

$$\delta(x) = x(0) = \int_{-\infty}^{\infty} F(t) x(t) \, dt$$

for all $x \in D$. Choose, for $a > 0$, $x(t) = \exp\{a^2(t^2 - a^2)^{-1}\}$ on $(-a, a)$ and $x(t) = 0$ otherwise. Then $x \in D$ and so

$$e^{-1} = \left| \int_{-a}^{a} F(t) x(t) \, dt \right| \leqslant e^{-1} \int_{-a}^{a} |F(t)| \, dt.$$

But it is a standard result in the theory of the Lebesgue integral that $\int_{-a}^{a} |F(t)| \, dt \to 0$ as $a \to 0$. Hence we have obtained a contradiction. It follows that δ is singular.

It is clear that D' is a linear space under the pointwise operations for functions. Convergence of sequences in D' is defined by saying that (f_k) converges in D' if and only if $(f_k(x))$ converges in C for every $x \in D$. If, for each $x \in D$, the limit of $(f_k(x))$ is $f(x)$, then f is linear, since $f(\lambda x + \mu y) = \lim_k f_k(\lambda x + \mu y) = \lim_k \{\lambda f_k(x) + \mu f_k(y)\} = \lambda f(x) + \mu f(y)$. It can be shown that f is necessarily continuous, i.e. $x_n \to 0$ (in D) implies $f(x_n) \to 0$. However, the proof is rather intricate, and those interested are referred to the book of Gelfand and Shilov.

A series of distributions $\sum_{k=1}^{\infty} f_k$ is said to converge to f if and only if $\sum_{k=1}^{\infty} f_k(x)$ converges to $f(x)$ for every $x \in D$. By our previous remarks the sum function of a convergent series of distributions is itself a distribution.

Example 17. Let $F_k(t) = \sin kt$ and suppose f_k is the regular distribution generated by F_k. Then $f_k \to \theta$ in D'. For, take $x \in D$, and suppose x vanishes outside $(-a, a)$. Then, on integration by parts,

$$|f_k(x)| = \left| \int_{-a}^{a} \sin kt . x(t) \, dt \right| = \left| \frac{1}{k} \int_{-a}^{a} \cos kt . x'(t) \, dt \right|$$

$$\leqslant \frac{1}{k} \int_{-a}^{a} |x'(t)| \, dt \to 0 \quad (k \to \infty).$$

Note, incidentally, that $(F_k(t))$ converges pointwise only at $t = n\pi$.

We shall now define the derivative of a distribution in such a way that every distribution has a derivative which is itself a distribution.

To motivate the definition consider a function $F(t)$ which has a continuous derivative on R. Then F generates a regular distribution f, and integration by parts gives

$$-\int_{-\infty}^{\infty} F(t)\, x'(t)\, dt = \int_{-\infty}^{\infty} F'(t)\, x(t)\, dt.$$

It is thus natural to define, in the general case, f' by $f'(x) = f(-x')$. Since $-x' \in D$ whenever $x \in D$, we see that f' is well-defined on D. Obviously f' is linear, and if $x_n \to 0$ (in D) then $-x'_n \to 0$ (in D) and so, by continuity of f, $f(-x'_n) \to 0$, i.e. $f'(x_n) \to 0$. Hence $f' \in D'$.

Example 18. (i) $\delta'(x) = \delta(-x') = -x'(0)$.

(ii) Consider the Heaviside unit step function 1_+, defined by $1_+(0) = \frac{1}{2}$, $1_+(t) = 1$ $(t > 0)$, $1_+(t) = 0$ $(t < 0)$. Then 1_+ generates the regular distribution $H(x) = \int_0^\infty x(t)\, dt$, called the Heaviside distribution. We then have $H' = \delta$. For

$$H'(x) = -\int_0^\infty x'(t)\, dt = -\left\{ \lim_{T \to \infty} x(T) - x(0) \right\}$$

$$= x(0) = \delta(x).$$

One of the main advantages of distribution theory is that any convergent series of distributions may be differentiated term by term. Of course this procedure is not valid in general in ordinary analysis.

Theorem 8. (i) *If $f_k \to f$ (in D') then $f'_k \to f'$ (in D').*
   (ii) *If $\Sigma g_k = g$ (in D') then $\Sigma g'_k = g'$ (in D').*

Proof. (i) $f'_k(x) = f_k(-x') \to f(-x')$, since $-x' \in D$ whenever $x \in D$. Hence $f'_k(x) \to f'(x)$ as $k \to \infty$ for each $x \in D$, i.e. $f'_k \to f'$ (in D').
   (ii) This follows on writing

$$f_k(x) = \sum_{n=1}^{k} g_n(x),$$

and using (i) and the fact that

$$f'_k(x) = \sum_{n=1}^{k} g'_n(x).$$

Example 19. Let $F_k(t) = (\sin kt)/\pi t$ and suppose f_k is the regular distribution corresponding to F_k. Then it is not hard to show that $f_k(x) \to x(0)$ $(k \to \infty)$, for each $x \in D$. Hence $f_k \to \delta$ (in D'). By theorem 8 (i) it follows that $f'_k \to \delta'$ (in D').

Exercises 5

1. Prove that the function $x(t)$ in example 12 is an element of D.
2. Construct a function in D which vanishes for $|t| \geqslant 2$ and has value 1 for $|t| \leqslant 1$.
3. Find which of the following define distributions: (i) $f(x) = \{x(0)\}^2$, (ii) $f(x) = \sum_{k=0}^{\infty} x^{(k)}(0)$, (iii) $f(x) = \sup \{x(t)|t \in R\}$.
4. Let $(F_k(t))$ be Lebesgue integrable on every finite interval and suppose that $F_k(t) \to F(t)$ almost everywhere on R. Suppose also that $|F_k(t)| \leqslant g(t)$, where g is Lebesgue integrable on R. Prove that $f_k \to f$ (in D'), where f_k, f are the regular distributions generated by F_k, F. As a hint, use Lebesgue's dominated convergence theorem (see Rudin's book).
5. Define $F_k(t) = k/2$ for $|t| < 1/k$, $F_k(t) = 0$ for $|t| \geqslant 1/k$. Prove that the sequence (f_k), of regular distributions generated by (F_k), converges to the delta distribution.
6. Do as in 5, with $F_k(t) = (k/\pi)^{\frac{1}{2}} e^{-kt^2}$. Prove also that $f_k(x) = x(0) + O(k^{-\frac{1}{2}})$, for each $x \in D$.
7. Do as in 5, with $F_k(t)$ of example 19. Sketch the graph of $F_k(t)$ for $k = 1, 3, 6$.

4

NORMED LINEAR SPACES

1. Convergence and completeness

In this chapter we are primarily concerned with normed linear spaces. A linear space was defined in chapter 3 and unless specific mention is made to the contrary we shall suppose in future that the space is complex linear, i.e. the scalar multipliers are complex numbers. A norm $\|.\|$ on a linear space X was also defined in chapter 3 and a normed linear space is just the pair $(X, \|.\|)$. To summarize:

Normed linear space

> *A normed linear space* $(X, \|.\|)$ *is a complex (or real) linear space* X *and a norm* $\|.\| : X \to R$ *such that* $\|x\| = 0$ *only if* $x = \theta$, $\|\lambda x\| = |\lambda| \|x\|$ *and* $\|x+y\| \leqslant \|x\| + \|y\|$.

We note that $\|\theta\| = 0$ and $\|x\| \geqslant 0$ for all x in X. There is a slight generalization of a normed linear space which sometimes arises. This is the p-normed space, in which the absolute homogeneity is altered slightly. Often we shall contract 'normed linear space' to 'normed space' in future.

p-normed space

> *Let* X *be a linear space,* $\|.\| : X \to R$ *and let* $p > 0$. *Then* $(X, \|.\|, p)$ *is a* p-*normed space if and only if* (i) $\|x\| = 0$ *only if* $x = \theta$, (ii) $\|\lambda x\| = |\lambda|^p \|x\|$, (iii) $\|x+y\| \leqslant \|x\| + \|y\|$.

A p-norm is necessarily non-negative and for economy we use the same notation for a p-norm and a 1-norm, i.e. a norm. No confusion will arise, since we shall explicitly state which sort of norm is being used.

If, in the definition above, we retain (ii) and (iii) but drop (i) then we have a p-seminormed space. A 1-seminormed space is then called a *seminormed space*.

Example 1. Let $0 < p < 1$. Then the space l_p of chapter 2 is a p-normed space, with p-norm

$$\|x\| = \Sigma |x_k|^p \quad (x \in l_p).$$

Of course $\|x\|$ is the norm for l_1 when $p = 1$.

The next theorem shows that a p-normed space is a special type of metric space.

Theorem 1. *Each p-normed linear space is a metric space with* $d(x, y) = \|x - y\|$ *and* $\|x\| = d(x, \theta)$. *The converse is not generally true.*

Proof. The first part is trivial. Now if (X, d) is a linear space which is also a metric space and we write $g(x) = d(x, \theta)$ then g is not generally a norm. For example, take $X = l(p)$ with $p_k = k^{-1}$ and

$$d(x, y) = \Sigma |x_k - y_k|^{p_k}.$$

We know from previous chapters that $l(p)$ is a linear metric space. Now g is not absolutely homogeneous, e.g. if $x = (0, 1, 0, 0, \ldots) \in l(p)$ then $g(2x) < 2g(x)$. Thus g is not a norm. The same type of argument shows that g is not a q-norm for any fixed $q > 0$.

Nearly all the examples of metric spaces given in chapter 2 are 'natural' normed spaces. Once the linear structure is introduced we put $\|x\| = d(x, \theta)$ and obtain a normed linear space. This does not work for all the examples—some exceptions being s, $l(p)$ (in general) and I.

Defining the algebraic operations of addition and scalar multiplication as in chapter 3 some explicit examples are

$$\|x\| = \sup |x_n| \text{ in } c \text{ and } l_\infty,$$
$$\|x\| = \max |x_n| \text{ in } c_0,$$
$$\|x\| = \max |x(t)| \text{ in } C[0, 1],$$
$$\|x\| = (\Sigma |x_k|^p)^{1/p} \text{ in } l_p \ (p \geqslant 1),$$
$$\|x\| = \max |x(t)| \text{ in } C(K), \ K \text{ compact.}$$

In a normed space, whenever we use topological concepts, it is to be understood that the topology is the metric topology generated by the norm. Thus for example $S(\theta, 1)$ is the unit sphere

$$\{x | d(x, \theta) < 1\} = \{x | \ \|x\| < 1\}.$$

Perhaps the most important type of normed space is that which is complete (as a metric space). In honour of Banach such a space is called a

Banach space

A Banach space X is a complete normed linear space. Completeness means that if $\|x_m - x_n\| \to 0$ $(m, n \to \infty)$, where $x_n \in X$, then there exists $x \in X$ such that $\|x_n - x\| \to 0$ $(n \to \infty)$.

Example 2. By our earlier examples we now see that R, C, l_∞, l_p $(p \geqslant 1)$, c, c_0 and $C[a, b]$ are Banach spaces.

Many results or definitions are however valid without completeness being assumed and we shall therefore use completeness only when necessary.

In a normed space convergence and absolute convergence of series may be naturally defined.

Series in normed spaces

Let X be a normed linear space. We say that the series

$$\Sigma x_n = \sum_{n=1}^{\infty} x_n,$$

with $x_n \in X$, converges to $s \in X$ if and only if

$$(s_n) = (x_1, x_1 + x_2, x_1 + x_2 + x_3, \ldots)$$

converges to s, i.e. $\|s_n - s\| \to 0$ $(n \to \infty)$. Then we write $\Sigma x_n = s$.

A series Σx_n is called absolutely convergent if and only if $\Sigma \|x_n\| < \infty$.

In R, every absolutely convergent series converges. This depends on completeness. The following gives a nice series characterization of a Banach space.

Theorem 2. *A normed linear space X is complete if and only if every absolutely convergent series is convergent.*

Proof. (i) Let X be complete and Σx_n absolutely convergent. Then for $n > m$, $\|s_n - s_m\| = \|x_{m+1} + \ldots + x_n\| \leqslant \|x_{m+1}\| + \ldots + \|x_n\| \to 0$ $(m \to \infty)$ by the general convergence principle for real series. Hence (s_n) is Cauchy and so convergent, i.e. Σx_n converges.

(ii) Let every absolutely convergent series be convergent. Let (x_n) be Cauchy in X. Then we can find $n_1 < n_2 < \ldots$ such that

$$\|x_{n_{k+1}} - x_{n_k}\| < 2^{-k} \quad (k = 1, 2, \ldots),$$

whence
$$\sum_{1}^{\infty} \|x_{n_{k+1}} - x_{n_k}\| < \infty.$$

By our assumption it follows that $\Sigma(x_{n_{k+1}} - x_{n_k})$ converges. This last series telescopes and we see that (x_{n_k}) converges. Hence the Cauchy sequence (x_n) has a convergent subsequence (x_{n_k}) and so the whole sequence (x_n) converges by theorem 2, section 2, chapter 2. Consequently, X is complete.

We note that the proof of theorem 2 does not use the absolute homogeneity of the norm. In fact the result will hold in the more general situation in which the norm is replaced by any non-negative subadditive function $g : X \to R$, though in this case the limits may not be unique.

As an example of the use of theorem 2 we consider the completeness of the Lebesgue integration space $L_p[0, 1]$.

The space $L_p[0, 1]$, $p > 1$. This is one of the classical Banach spaces. The elements of L_p are functions x such that $|x|^p$ is Lebesgue integrable on $[0, 1]$, i.e. $x \in L_p$ if and only if

$$\int_0^1 |x(t)|^p \, dt < \infty.$$

The natural seminorm in L_p is

$$\|x\| = \left(\int_0^1 |x(t)|^p \, dt \right)^{1/p}.$$

This is not a norm, since $\|x\| = 0$ implies only that $x(t) = 0$ almost everywhere on $[0, 1]$. However, by considering equivalence classes as in example 6, chapter 2, we may, if we wish, construct a normed space from L_p. The triangle inequality $\|x+y\| \leqslant \|x\| + \|y\|$ is in this case the Minkowski inequality for Lebesgue integrals.

To prove the completeness of L_p in the seminorm given above we need a fundamental theorem of Lebesgue integration, known as Lebesgue's monotone convergence theorem. We state this below, with two of its immediate corollaries. Proofs may be found, for example, in the book *Principles of Mathematical Analysis*, by W. Rudin.

Lebesgue's monotone convergence theorem. If (x_n) is a sequence of functions, measurable on $[0, 1]$, such that $0 \leqslant x_1(t) \leqslant x_2(t) \leqslant \ldots$ and $\lim_n x_n(t) = x(t)$ on $[0, 1]$, then

$$\lim_n \int_0^1 x_n(t) \, dt = \int_0^1 x(t) \, dt.$$

Corollary 1. *If (x_n) is a sequence of non-negative functions, measurable on $[0, 1]$ and*

$$x(t) = \sum_1^\infty x_n(t),$$

then

$$\int_0^1 x(t)\, dt = \sum_1^\infty \int_0^1 x_n(t)\, dt.$$

Corollary 2 (Fatou's lemma). *If (x_n) is a sequence of non-negative functions, measurable on $[0, 1]$ and $x(t) = \lim \inf_n x_n(t)$ on $[0, 1]$, then*

$$\int_0^1 x(t)\, dt \leqslant \lim \inf_n \int_0^1 x_n(t)\, dt.$$

Now we are in a position to apply theorem 2 to show that L_p is complete. Take any absolutely convergent series Σx_k, where $x_k \in L_p$ $(k = 1, 2, \ldots)$. We shall show that there exists $s \in L_p$ such that $s = \Sigma x_k$, convergence being in the seminorm of L_p. Our hypothesis is that $\Sigma \|x_k\| < \infty$, whence there exists N such that

$$\sum_{k=m+1}^\infty \|x_k\| < \epsilon \quad (m > N).$$

By Hölder's inequality (for integrals),

$$\int_0^1 |x_k(t)|\, dt \leqslant \|x_k\| \left(\int_0^1 1 . dt \right)^{1/q} = \|x_k\|,$$

and so, by corollary 1, above,

$$\int \Sigma |x_k(t)|\, dt = \Sigma \int |x_k(t)|\, dt \leqslant \Sigma \|x_k\| < \infty.$$

It follows that $\Sigma |x_k(t)| < \infty$, almost everywhere on $[0, 1]$. For if $\Sigma |x_k(t)| = \infty$ on a set of positive measure, then $\int \Sigma |x_k(t)|\, dt = \infty$. Hence, $\Sigma x_k(t)$ converges almost everywhere, i.e. there exists

$$\lim_n \sum_{k=1}^n x_k(t) = s(t), \quad \text{say,}$$

almost everywhere. Now take $m > N$. Then

$$\left\| s - \sum_1^m x_k \right\|^p = \int \left| s(t) - \sum_1^m x_k(t) \right|^p dt$$

$$= \int \lim_n \left| \sum_1^n x_k(t) - \sum_1^m x_k(t) \right|^p dt$$

$$\leqslant \lim \inf_n \int \left| \sum_{m+1}^n x_k(t) \right|^p dt$$

by corollary 2. But

$$\int \left| \sum_{m+1}^{n} x_k(t) \right|^p dt = \left\| \sum_{m+1}^{n} x_k \right\|^p$$

and so

$$\left\| s - \sum_{1}^{m} x_k \right\|^p \leqslant \lim\inf_n \left(\sum_{m+1}^{n} \|x_k\| \right)^p < \epsilon^p.$$

This means that

$$\sum_{1}^{m} x_k \to s$$

in the sense of seminorm convergence in L_p. Finally, we have, for $m > N$,

$$\|s\| \leqslant \left\| s - \sum_{1}^{m} x_k \right\| + \sum_{1}^{m} \|x_k\| < \epsilon + \sum_{1}^{\infty} \|x_k\| < \infty,$$

whence $s \in L_p$. Consequently L_p is complete.

The case $p = 2$ of this completeness result is important in the theory of Fourier series, in connection with the Riesz–Fischer theorem. For details of this we refer to the later chapter (chapter 6) on Hilbert space.

If we are given a normed space X we may always construct other normed spaces from it. These spaces are called

Normed factorspaces

Let X be a normed space and M a closed subspace of X. Define $x_1 \equiv x_2 \pmod{M}$ to mean that $x_1 - x_2 \in M$. Then $\equiv$ defines an equivalence relation and X/M, the factorspace X modulo M, is the set of all equivalence classes E_x. The norm in X/M is defined by

$$\|E_x\| = \inf\{\|y\| \,|\, y \in E_x\}. \tag{1}$$

As was pointed out in chapter 3, X/M is a linear space with the definitions $E_x + E_y = E_{x+y}$, $\lambda E_x = E_{\lambda x}$. Now we check that (1) is a norm on X/M. As samples we show that $\|E\| = 0$ implies $E = M$, the zero in X/M, and also we check the triangle inequality.

Let $\|E\| = 0$. Then there exists a sequence $(x_n) \in E$ such that $x_n \to \theta$. Since M is closed it is clear that E is closed and so $\theta \in E$, whence $E = E_\theta = M$.

Now let E, F be elements of X/M. By (1), given any $\epsilon > 0$, there exist $x \in E$, $y \in F$ such that

$$\|x\| < \|E\| + \epsilon, \quad \|y\| < \|F\| + \epsilon,$$

whence

$$\|x + y\| \leqslant \|x\| + \|y\| < \|E\| + \|F\| + 2\epsilon.$$

Now $x+y\in E+F$, so that $\|E+F\| \leqslant \|x+y\|$. Therefore

$$\|E+F\| \leqslant \|E\| + \|F\|,$$

since ϵ is arbitrary.

Provided X is a Banach space the normed factorspace X/M is also a Banach space.

Theorem 3. *If X is a Banach space and M a closed subspace of X, then X/M is also a Banach space, under the norm defined above.*

Proof. We shall use theorem 2 and prove that if $\Sigma \|E(k)\| < \infty$ then $\Sigma E(k)$ converges, where $E(k)\in X/M$ for $k = 1, 2, \dots$. Now by definition of $\|E(k)\|$ there exists $x_k\in E(k)$ such that

$$\|x_k\| < \|E(k)\| + 2^{-k} \quad (k = 1, 2, \dots).$$

Hence $\Sigma \|x_k\| < 1+\Sigma \|E(k)\| < \infty$ and the completeness of X implies, by theorem 2, that $\Sigma x_k = x$, say. Let E_x be the equivalence class containing this x. Then we shall show that $\Sigma E(k) = E_x$, i.e. that

$$s_n = \sum_{k=1}^{n} E(k) \to E_x \quad (n\to\infty).$$

For we have

$$s_n - E_x = E(1)+\dots+E(n) - E_x = E_{x_1}+\dots+E_{x_n} - E_x = E_{x_1+\dots+x_n-x}.$$

Hence, by (1),

$$\|s_n - E_x\| \leqslant \|x_1+\dots+x_n - x\| \to 0 \quad (n\to\infty),$$

since $\Sigma x_n = x$. Thus, every absolutely convergent series in X/M converges, i.e. X/M is complete. This proves the theorem.

Exercises 1

1. In connection with normed factorspaces prove that each equivalence class E_x is closed when M is closed. Prove also that $\|\lambda E\| = |\lambda| \|E\|$, with the definition of $\|E\|$ given in section 1.

2. In l_2 let M be the set of all sequences $x = (x_1, \dots, x_n, 0, 0, \dots)$ which have only a finite number of nonzero terms. Prove that M is a linear subspace of l_2 which is not closed. As a hint consider the sequence $x_1 = (1, 0, 0, \dots)$, $x_2 = (1, \frac{1}{2}, 0, 0, \dots)$, $x_3 = (1, \frac{1}{2}, \frac{1}{4}, 0, 0, \dots)$, ... in M.

3. In a normed space show that
   (i) $\|x-y\| \geqslant |\,\|x\|-\|y\|\,|$,
   (ii) $x_n\to x$ implies $\|x_n\| \to \|x\|$,
   (iii) $\lambda_n\to\lambda, \mu_n\to\mu \text{ (in } C\text{)}$ and $x_n\to x, y_n\to y$ imply $\lambda_n x_n+\mu_n y_n\to\lambda x+\mu y$,
   (iv) $x_n\to x$ implies $x_n/\|x_n\| \to x/\|x\|$, with obvious restrictions on x_n and x.

4. Let X be a Banach space and suppose the normed space Y is isometric to X. Prove that Y is a Banach space.

5. Show that l_p $(0 < p < 1)$ is a complete p-normed space.

6. Let S denote the set of all real continuous functions $x = x(t)$ on $[0, 1]$. Show that
$$\|x\| = \left(\int_0^1 x^2(t)\, dt \right)^{\frac{1}{2}}$$

is a norm on S, but that S is not complete under this norm. As a hint consider $x_n(t) = 1$ on $[0, \frac{1}{2}]$, $= 0$ on $[\frac{1}{2}+\frac{1}{2}n, 1]$, completing the definition of x_n by drawing a straight line from the point $(\frac{1}{2}, 1)$ to the point $(\frac{1}{2}+\frac{1}{2}n, 0)$.

7. In a normed space prove that $\|\Sigma x_n\| \leqslant \Sigma \|x_n\|$ whenever Σx_n converges.

8. Let $\Sigma(x_n - x_{n+1})$ be absolutely convergent in a normed space. Is (x_n) Cauchy? Is (x_n) convergent?

9. Let $e_1 = (1, 0, 0, \ldots)$, $e_2 = (0, 1, 0, \ldots)$, $e_3 = (0, 0, 1, 0, \ldots), \ldots$ Show that

$$\Sigma \frac{e_k}{k \log (k+1)}$$

is convergent but not absolutely convergent in c_0. Find the sum of the series.

10. M is a subspace of a normed space. Prove that the closure $\bar{M}$ of M is a closed subspace.

11. If S is a dense subset of a normed space X, show that every element of X can be written as the sum of an absolutely convergent series whose terms are from S.

12. For any $p > 0$, w_p is the set of all sequences $x = (x_k)$ such that there is a number l for which

$$\frac{1}{n} \sum_{k=1}^n |x_k - l|^p \to 0 \quad (n \to \infty).$$

Prove that if $x \in w_p$ then l is unique. Show also that w_p is a linear space with the usual operations $\lambda(x_k) = (\lambda x_k)$, $(x_k) + (y_k) = (x_k + y_k)$.
   Define

$$\|x\| = \sup_n \left(n^{-1} \sum_{k=1}^n |x_k|^p \right)^{1/p} \quad (1 \leqslant p < \infty),$$

$$\|x\| = \sup_n n^{-1} \sum_{k=1}^n |x_k|^p \quad (0 < p < 1).$$

Show that $\|x\|$ is a norm when $p \geqslant 1$ and a p-norm when $0 < p < 1$. Prove also that w_p is a Banach space when $p \geqslant 1$ and a complete p-normed space when $0 < p < 1$.
   We shall be examining some further properties of the space w_p in chapter 7.

13. Let $X \neq \{\theta\}$ be a normed space. Prove that X is a Banach space if and only if $\{x \in X \mid \|x\| = 1\}$ is complete.

2. Linear operators and functionals

In many problems in analysis one finds that one is considering linear operators on some linear space which have values in another linear space (possibly the same space). For example, in the theory of integral equations, operators A of the type defined by

$$(Ax)(s) = \int_0^1 a(s,t)\,x(t)\,dt$$

occur. In this equation we consider $a(s,t)$ as a given function, continuous on $[0,1] \times [0,1]$ and $x = x(t)$ continuous on $[0,1]$. Then it is clear that Ax is a continuous function of $s \in [0,1]$, so that

$$A : C[0,1] \to C[0,1].$$

Because the function A operates on each x, sending it into another continuous function, we may reasonably call A an operator, as is conventional in functional analysis. By the 'linear' properties of the integral one sees that A is a linear operator, in the sense that

$$A(\lambda x + \mu y) = \lambda A(x) + \mu A(y),$$

for scalars λ, μ and functions $x, y \in C[0,1]$.

Numerous special types of linear operator will be considered in this and later chapters (in chapter 7 we study linear operators defined by infinite matrices). For the moment we shall be concerned with some general definitions and properties.

Linear operator

Let X, Y be linear spaces. Then a function $A : X \to Y$ is called a linear operator (or map, transformation) if and only if, for all $x_1, x_2 \in X$, and all scalars λ, μ,

$$A(\lambda x_1 + \mu x_2) = \lambda A(x_1) + \mu A(x_2). \tag{2}$$

Linear functional

A is a linear functional on X if $A : X \to C$ is a linear operator, i.e. a linear functional is a complex-valued linear operator.

Kernel of an operator

The kernel of an operator A is defined as
$$\mathrm{Ker}\,(A) = \{x \in X \mid A(x) = \theta\}.$$
The kernel of a linear operator is clearly a subspace of X.

Note. We are usually supposing that X is complex linear. If X is real linear then a linear functional is defined as a real-valued linear operator on X.

Example 3. (i) Let I be the space of all integral functions f on C. Then the differential operator $A : I \to I$ defined by $A(f) = f'$ is a linear operator on I into itself. The linearity expresses familiar properties of the derivative but it is nontrivial that f' is an integral function. That this is the case follows from the well-known theorem of complex variable on the indefinite differentiability of functions analytic on a domain (see Ahlfors, *Complex Analysis*).

(ii) If c is the space of convergent sequences $x = (x_n)$, then $A(x) = \lim x_n$ is a linear functional on c. Another example of a linear functional on c is $B : c \to C$, given by

$$B(x) = \Sigma n^{-2} x_n,$$

the series being absolutely convergent for each x in c.

(iii) A general linear operator $A : X \to Y$ has the property that $A(\theta) = \theta$, as is obvious on putting $\lambda = \mu = 0$ in (2). No confusion should arise from the fact that we shall not usually distinguish between the zero in X and that in Y.

Operators on normed spaces, which are linear and continuous are of especial interest. The continuity is understood to be the metric continuity given by the norm. Thus $A : X \to Y$ is continuous at $x_0 \in X$ if for every $\epsilon > 0$ there exists $\delta(x_0, \epsilon)$ such that $\|A(x) - A(x_0)\| < \epsilon$ whenever $\|x - x_0\| < \delta$. Here and in future we shall not usually bother to distinguish norms in X from those in Y.

Another type of operator on a normed space, which actually turns out to be the same thing as a continuous linear operator is a

Bounded linear operator

A linear operator $A : X \to Y$ is called bounded if and only if there exists a constant M such that

$$\|A(x)\| \leqslant M \|x\| \quad \text{for all} \quad x \in X.$$

One should observe that this definition of bounded is not the same as that for an ordinary complex function: $|f(z)| \leqslant M$ for all z in some set. Note also that a bounded functional on X satisfies $|f(x)| \leqslant M \|x\|$, for all x in X.

Example 4. The operators A and B of example 3 (ii) are bounded linear functionals on c. For, with $\|x\| = \sup |x_n|$, we have $|x_n| \leqslant \|x\|$ for all n and so $|A(x)| = \lim |x_n| \leqslant \|x\|$ on c. Also,

$$|B(x)| \leqslant \Sigma n^{-2} |x_n| \leqslant \frac{\pi^2}{6} \|x\|.$$

Now we give some properties of continuous linear operators.

Theorem 4. *Let X, Y be normed spaces and $A : X \to Y$ a linear operator. Then*

     (i)   *A continuous at the origin implies A uniformly continuous on X.*

     (ii)   *A is continuous on X if and only if it is bounded.*

Proof. (i) We have $\|A(x)\| < \epsilon$ whenever $\|x\| < \delta$. Hence, if $\|x - y\| < \delta$ then $\|A(x - y)\| = \|A(x) - A(y)\| < \epsilon$, by linearity of A.

     (ii) First let A be bounded: $\|A(x)\| \leqslant M \|x\|$ on X. Then

$$\|A(x) - A(y)\| = \|A(x - y)\| \leqslant M \|x - y\| < \epsilon,$$

if $\|x - y\| < \epsilon/M$. Hence A is uniformly continuous.

Now suppose that A is continuous on X. Then it is continuous at θ and so there exists $\delta = \delta(1)$ such that $\|A(x)\| < 1$ whenever $\|x\| < \delta$. Take any $x \neq \theta$. Then

$$\left\| \frac{\delta x}{2 \|x\|} \right\| = \frac{\delta}{2}$$

and so
$$\left\| A\left(\frac{\delta x}{2 \|x\|} \right) \right\| < 1, \quad \|A(x)\| < \frac{2}{\delta} \|x\|.$$

If $x = \theta$ then $\|A(x)\| = 0$ and so $\|A(x)\| \leqslant 2\delta^{-1} \|x\|$ on X, i.e. A is bounded.

There is a result analogous to theorem 4 (i) for seminorms.

Theorem 5. *Let p be a seminorm on X. Then p is continuous on X if and only if it is continuous at the origin. Also, this is equivalent to the boundedness of p, i.e. $p(x) \leqslant M \|x\|$ on X for some M.*

Proof. If p is continuous on X then of course it is continuous at the origin. Conversely, if p is continuous at the origin then we use the argument of theorem 4 (ii) and prove that $p(x) \leqslant 2\delta^{-1} \|x\|$ on X. From this we get, since $|p(x) - p(y)| \leqslant p(x - y)$, that p is actually uniformly continuous on X.

We now know that a continuous linear operator on a normed space is the same thing as a bounded linear operator. Another way of describing continuous linear functionals is to look at the kernel.

Theorem 6. *Let $A : X \to C$ be linear. Then A is continuous if and only if* Ker (A) *is closed.*

Proof. (i) Let A be continuous. Then, since Ker $(A) = A^{-1}(\{0\})$, we see that Ker (A) is closed. For it is the inverse image of the closed set $\{0\}$.

Another way of proving this is to let $x \in \overline{\text{Ker}(A)}$. Then there exists $(x_n) \in \text{Ker}(A)$ such that $x_n \to x$. Hence, since A is continuous and $A(x_n) = 0$ we have $A(x) = 0$, i.e. $x \in \text{Ker}(A)$, whence Ker (A) is closed.

We note that the argument will work if C is replaced by any normed space Y.

(ii) Now let Ker (A) be a closed set. If $A \equiv 0$ then A is continuous. Suppose $A \not\equiv 0$, so that $\sim$ Ker (A) is nonempty and open. Now there exists a in $\sim$ Ker (A) such that $A(a) \neq 0$. Hence $A(b) = 1$, where $b = a/A(a) \in \sim$ Ker (A). Consequently there exists $S(b, r) \subset \sim$ Ker (A).

We shall shortly show that

$$S(\theta, r) \subset V = \{x \mid |A(x)| < 1\}. \tag{3}$$

This will imply $|A(x)| < 1$ whenever $\|x\| < r$ and so, by the argument of theorem 4 (ii), we get $|A(x)| \leqslant 2r^{-1} \|x\|$ on X, i.e. A is continuous on X.

To prove (3) take $x \in S(\theta, r)$ and suppose $x \notin V$. Then

$$y = -x/A(x) \in S(\theta, r) \quad \text{and} \quad A(b+y) = 1 + A(y) = 0.$$

Also, $b + y \in S(b, r)$, since $\|y\| < r$. Combining the last parts we get $b + y \in \text{Ker}(A)$ and $b + y \in S(b, r)$, which means that

$$\text{Ker}(A) \cap S(b, r) \neq \varnothing.$$

But this contradicts the fact that $S(b, r) \subset \sim$ Ker (A). Hence $x \in S(\theta, r)$ implies $x \in V$, which proves (3) and so the theorem (see the comment after (3)).

Before we proceed further we fix some notation and terminology.

$L(X, Y)$. Let X, Y be linear spaces. Then $L(X, Y)$ denotes the set of all linear operators on X into Y.

$X^\dagger$. This is $L(X, C)$, the set of all linear functionals on X. It is usual to call $X^\dagger$ the algebraic dual of X.

$B(X, Y)$. Let X, Y be normed spaces. Then $B(X, Y)$ denotes the set of all bounded (i.e. continuous) linear operators on X into Y.

X^*. This is $B(X, C)$, the set of all bounded linear functionals on X. We call X^* the dual (or continuous dual) of X.

For any normed X, Y it is in the nature of things that

$$B(X, Y) \subset L(X, Y).$$

In general the inclusion is strict, as the next example shows.

Example 5. $X^* \subset X^\dagger$ strictly, where $X = l_1$ with the norm of l_∞. This can be shown by considering

$$f(x) = \Sigma x_k \quad (x \in l_1).$$

It is obvious that $f : l_1 \to C$ and that f is linear on l_1. However, f is not bounded. For suppose it were. Then there exists an integer N such that $|f(x)| \leqslant N \|x\|$ for all x in l_1. Let $y \in l_1$ be defined by

$$y_k = 1 \quad (1 \leqslant k \leqslant N+1), \qquad y_k = 0 \quad (k > N+1).$$

Then $\|y\| = \sup |y_k| = 1$, $f(y) = N+1$, and by our supposition we get $N+1 \leqslant N$, a contradiction. Hence f is not bounded.

We may note that the normed space X of example 5 is *infinite dimensional*. It turns out that for *finite* dimensional spaces it is always true that $X^* = X^\dagger$, i.e. every linear functional on X is necessarily continuous (see exercises 2, question 10).

The sets $L(X, Y)$ and $B(X, Y)$ defined above are sets of functions (operators) whose values lie in the linear space Y. It is natural therefore to make these sets into linear spaces. This we do by defining $\lambda A + \mu B$, with $A, B \in L(X, Y)$, by

$$(\lambda A + \mu B)(x) = \lambda A(x) + \mu B(x).$$

It is immediate that $\lambda A + \mu B \in L(X, Y)$ and that $L(X, Y)$ is thus a linear space. Also, $B(X, Y)$ is easily shown to be a linear subspace of $L(X, Y)$.

To make $B(X, Y)$ into a normed space we must introduce the norm of an element $A \in B(X, Y)$.

Norm of a bounded operator

Let $A \in B(X, Y)$. Then the norm of A is defined as

$$\|A\| = \sup_{x \neq \theta} \frac{\|A(x)\|}{\|x\|} < \infty.$$

That the supremum is finite follows from the fact that

$$\|A(x)\| \leqslant M \|x\| \quad \text{when} \quad A \in B(X, Y).$$

Now we show that $\|A\|$ is indeed a norm on $B(X, Y)$ and also we prove that $B(X, Y)$ is a Banach space when Y is a Banach space, irrespective of whether X is a Banach space.

Theorem 7. (i) *The linear space $B(X, Y)$ of all bounded linear operators on the normed space X into the normed space Y is a normed space with $\|A\| = \sup \{\|A(x)\|/\|x\| \,|x \neq \theta\}; A \in B(X, Y)$.*

(ii) *If Y is a Banach space then $B(X, Y)$ is a Banach space under the norm of* (i).

Proof. (i) If $A_1, A_2 \in B(X, Y)$ then

$$\|A_1(x)\| \leqslant \|A_1\| \, \|x\| \quad \text{and} \quad \|A_2(x)\| \leqslant \|A_2\| \, \|x\|$$

for all $x \in X$. Hence

$$\|(A_1 + A_2)(x)\| = \|A_1(x) + A_2(x)\| \leqslant (\|A_1\| + \|A_2\|) \, \|x\|$$

for all $x \in X$, which implies that $\|A_1 + A_2\| \leqslant \|A_1\| + \|A_2\|$. It is trivial that $\|\lambda A\| = |\lambda| \, \|A\|$ and so it remains to show that $\|A\| = 0$ implies $A \equiv \theta$. But if $\|A\| = 0$ then we deduce from $\|A(x)\| \leqslant \|A\| \, \|x\|$ that $A(x) = \theta$ on X, i.e. $A \equiv \theta$.

(ii) Let (A_n) be Cauchy in $B(X, Y)$. Then for each $x \in X$,

$$\|A_n(x) - A_m(x)\| \leqslant \|A_n - A_m\| \cdot \|x\| \to 0 \quad (n, m \to \infty),$$

so that $(A_n(x))$ is Cauchy in Y, whence there exists $\lim_n A_n(x) = A(x)$, say. We want to show that $A \in B(X, Y)$ and that $A_n \to A$ in the norm of $B(X, Y)$. Since A_n is linear we have $A_n(\lambda x + \mu x') = \lambda A_n(x) + \mu A_n(x')$ and so, letting $n \to \infty$, we obtain $A(\lambda x + \mu x') = \lambda A(x) + \mu A(x')$, whence $A \in L(X, Y)$. Now A is bounded since

$$\|A(x)\| = \lim_n \|A_n(x)\| \leqslant \|x\| + \|A_N\| \, \|x\|,$$

where $N = N(1)$ comes from the Cauchy condition on (A_n).

Finally, by definition of $\|A_n - A\|$ as a supremum, we have for every $\epsilon > 0$ an $x \neq \theta$ such that

$$\|A_n - A\| < \|A_n(x) - A(x)\|/\|x\| + \epsilon/2$$

$$= \|A_n(x/\|x\|) - A(x/\|x\|)\| + \epsilon/2$$

$$< \epsilon \quad \text{if} \quad n > N(\epsilon),$$

on using the fact that $A_n(x/\|x\|) \to A(x/\|x\|)$ in the norm of Y.

Corollary. *The dual X^* of a normed space X, is always a Banach space under the norm*

$$\|f\| = \sup\{|f(x)|/\|x\|\ |x \ne \theta\}.$$

Proof. Put $Y = C$ in the theorem and recall that C is complete with the modulus norm.

From the general point of view, theorem 7 and its corollary tell us much of what we might reasonably expect to know about $B(X, Y)$ and X^*. However, if a normed space X is given in concrete form, e.g. c, l_∞ or $C[0, 1]$, then it is especially interesting to give an explicit characterization of the dual space X^*. This is not always easy to do and in fact, with the theory at our disposal, it is possible to carry this out only for one of the examples cited, viz. c. To characterize the dual of l_∞ one needs to know some measure theory (see the remarks which follow the table of dual spaces below). The dual of $C[0, 1]$ will be discussed after we have proved the Hahn–Banach extension theorem (see section 5 below).

Before we characterize c^* we introduce the idea of equivalent normed spaces which is helpful for dual characterization in general.

Equivalent normed spaces

Normed spaces X and Y are called equivalent if and only if they are isometrically isomorphic. This means that there exists a linear isometry $T : X \to Y$.

Note. If T is a linear isometry then $\|T(x)\| = \|x\|$ on X, since $T(\theta) = \theta$. Conversely, if T is linear and $\|T(x)\| = \|x\|$ on X, then

$$\|T(x) - T(y)\| = \|T(x-y)\| = \|x-y\|$$

so that T is an isometry, provided it is surjective. Thus X and Y are equivalent if and only if there is a linear surjective norm preserving mapping $T : X \to Y$.

It is clear that if $X \sim Y$ means X and Y are equivalent normed spaces, then $\sim$ is an equivalence relation on the class of all normed spaces, so that X and Y may be regarded as indistinguishable from the point of view of the theory of normed spaces. When $X \sim Y$ we feel free to 'identify' X and Y and sometimes even say that X *is* Y, which does no harm as long as we bear in mind what is intended.

Now we show that $c^* \sim l_1$, and in view of the preceding paragraph we say that the dual of c is l_1.

Theorem 8. (i) *If $f \in c^*$ then there is a number a and a sequence $(a_n) \in l_1$ such that, for all $x \in c$,*

$$f(x) = a \lim x_n + \Sigma a_n x_n. \tag{4}$$

Also the c^*-norm of f is

$$\|f\| = |a| + \Sigma |a_n|.$$

Conversely, if a and $(a_n) \in l_1$ are given, then the right side of (4) defines an element of c^*.

(ii) *c^* is equivalent to l_1.*

Proof. (i) The converse part is trivial. Suppose then that $f \in c^*$. Now from chapter 3 we know that $\{e, e_1, e_2, ...\}$ is a basis for c:

$$x = le + \Sigma(x_n - l)e_n,$$

where $x = (x_n) \in c$, $l = \lim x_n$, $e = (1, 1, 1, ...)$, $e_1 = (1, 0, 0, ...)$, ... By linearity and continuity of f we get

$$f(x) = lf(e) + \Sigma(x_n - l)f(e_n) \tag{5}$$

for all $x \in c$. Now take any $r \geqslant 1$ and let‡ $x_n = \operatorname{sgn} f(e_n)$ for $1 \leqslant n \leqslant r$, $x_n = 0$ for $n > r$. Then $x \in c_0$, $\|x\| = 1$ and so

$$|f(x)| = \sum_{n=1}^{r} |f(e_n)| \leqslant \|f\|, \tag{6}$$

since $|f(x)| \leqslant \|f\| \|x\|$ on c. It follows from (6) that

$$\Sigma |f(e_n)| = \sup_r \sum_{n=1}^{r} |f(e_n)| \leqslant \|f\| < \infty.$$

We now write (5) as $\qquad f(x) = al + \Sigma a_n x_n, \tag{7}$

where $a = f(e) - \Sigma f(e_n)$, $a_n = f(e_n)$, the series $\Sigma f(e_n)$ being absolutely convergent. From (7) we have, since $|\lim x_n| \leqslant \|x\|$,

$$|f(x)| \leqslant (|a| + \Sigma |a_n|) \|x\|,$$

whence $\|f\| \leqslant |a| + \Sigma |a_n|$. Also, for $\|x\| = 1$, we have $|f(x)| \leqslant \|f\|$, so we define, for any $r \geqslant 1$,

$$x_n = \operatorname{sgn} a_n \quad (1 \leqslant n \leqslant r),$$
$$= \operatorname{sgn} a \quad (n > r).$$

Then $x \in c$, $\|x\| = 1$, $\lim x_n = \operatorname{sgn} a$ and so

$$|f(x)| = \left| |a| + \sum_{n=1}^{r} |a_n| + \sum_{n=r+1}^{\infty} a_n \operatorname{sgn} a \right| \leqslant \|f\|.$$

‡ For complex z we define $\operatorname{sgn} z = |z|/z$ ($z \neq 0$) and $\operatorname{sgn} 0 = 1$. Then, for all z, we have $|\operatorname{sgn} z| = 1$ and $z \operatorname{sgn} z = |z|$.

Since $(a_n)\in l_1$ we have $\sum\limits_{n=r+1}^{\infty} a_n \to 0\ (r\to\infty)$, whence, letting $r\to\infty$ in the last inequality we get

$$|a|+\Sigma|a_n| \leqslant \|f\|.$$

Combining our results we see that $\|f\| = |a|+\Sigma|a_n|$. This proves (i).

It is easy to see that the representation (4) is unique. This uniqueness is required in part (ii).

(ii) Let $T:c^*\to l_1$ be defined by

$$T(f) = (a, a_1, a_2, \ldots),$$

where a, a_n appear from the representation in (i). By (i) we have $\|T(f)\| = |a|+|a_1|+|a_2|+\ldots = \|f\|$, $\|T(f)\|$ being the l_1 norm. Thus T is norm preserving. It is surjective by the 'conversely' part of (i). Finally, T is obviously linear, e.g. if $f\in c^*$ then

$$T(\lambda f) = (\lambda a, \lambda a_1, \ldots) = \lambda(a, a_1, \ldots) = \lambda T(f).$$

Additivity of T is similar.

Below we give a table which shows the duals of some of the spaces we have so far introduced. Some comments on the table are given after it.

TABLE OF DUAL SPACES

Space	Representation	Norm	Dual
c	$a \lim x_n + \Sigma a_n x_n$	$\|a\|+\Sigma\|a_n\|$	l_1
c_0	$\Sigma a_n x_n$	$\Sigma\|a_n\|$	l_1
$l_p\ (0 < p \leqslant 1)$	$\Sigma a_n x_n$	$\sup_n\|a_n\|$	l_∞
$l_p\ (1 < p < \infty)$	$\Sigma a_n x_n$	$(\Sigma\|a_n\|^q)^{1/q}$	l_q
l_∞	$\int_N x(k)\,d\mu$	$\int\|d\mu\|$	$\mathcal{M}$
$C[0,1]$	$\int_0^1 x(t)\,dg(t)$	$\int_0^1\|dg(t)\|$	$\widehat{BV}[0,1]$

In the table the space in the first column is the space whose dual is to be given. The representation, of the second column, means the form taken by each element of the dual; x being an element of the space. The third column gives the norm of the bounded linear functional and the final column refers to the 'identified' space, e.g. $c^* \sim l_1$, so the dual of c is given as l_1.

We have already proved the result concerning c and the results for c_0, l_p may be proved by similar methods. In connection with l_p^* the relation between p and q is $1/p+1/q = 1$. Note of course that

l_p $(0 < p < 1)$ is p-normed not normed and see question 3 of exercises 2 for the definition of operator norm in a p-normed space. The details for the space $C[0, 1]$ are given after we have proved the Hahn–Banach extension theorem (section 5 below). $\widehat{BV}[0, 1]$ denotes a certain subspace of the space of functions of bounded variation on $[0, 1]$.

Without being led too far afield we cannot prove the statements about l_∞. Its dual $\mathscr{M}$ is actually the space of bounded finitely additive set functions (or measures) μ defined on subsets of the set of positive integers N. It is to be remarked that the dual of the sequence space l_∞ is not a sequence space, though the duals of the other sequence spaces in the table are. This can be attributed to the fact that l_∞ has no basis (it is not separable) whereas the other spaces have.

We finish this section with some results of a slightly geometrical flavour. First we define a

Hyperplane through the origin

Let X be a linear space. A hyperplane H, through θ, is defined to be a maximal proper subspace of X, i.e. $H \neq X$ and if for any subspace M of X we have $H \subset M \subset X$, then either $M = H$ or $M = X$.

Hyperplane

A hyperplane is a set of the form $x_0 + H$, where H is a hyperplane through θ.

The kernel of a non-identically zero linear functional is a hyperplane through θ:

Theorem 9. (i) *If $f \in X^\dagger$, $f \not\equiv 0$, then $\mathrm{Ker}\,(f)$ is a hyperplane.*

(ii) *If $f \in X^*$, $f \not\equiv 0$, then $\mathrm{Ker}\,(f)$ is a closed hyperplane.*

Proof. We know by theorem 6 that $\mathrm{Ker}\,(f)$ is closed when $f \in X^*$. Hence we prove part (i) only. Now $f \not\equiv 0$ implies $\mathrm{Ker}\,(f) \neq X$ and $\mathrm{Ker}\,(f)$ is a subspace since $f(\lambda x + \mu y) = 0$ whenever $x, y \in \mathrm{Ker}\,(f)$. Suppose then that $\mathrm{Ker}\,(f) \subset M$, strictly, where M is a subspace of X. We want to show that $M = X$. By our supposition there exists

$$x_0 \in M \sim \mathrm{Ker}\,(f),$$

so that $f(x_0) \neq 0$. Take $x \in X$ and set

$$y = x - f(x)\,x_0/f(x_0).$$

Then $f(y) = 0$, whence $y \in \operatorname{Ker}(f) \subset M$, and so x, being a linear combination of elements of M, is in M. Thus $X \subset M$ and so $X = M$. Consequently, $\operatorname{Ker}(f)$ is a hyperplane.

Corollary. *If $f \in X^\dagger$, $f \not\equiv 0$, then $\{x \mid f(x) = \alpha\}$, α a fixed scalar, is a hyperplane.*

Proof. Let $S = \{x \mid f(x) = \alpha\}$. Since $f \not\equiv 0$ there exists x_1 such that $f(x_1) \neq 0$. Put $x_0 = \alpha x_1 / f(x_1)$. Then it is easy to check that

$$S = x_0 + \operatorname{Ker}(f).$$

Hence S is a hyperplane.

Example 6. Let $X = R^3$ and $f(x) = x_1 + x_2 + x_3$, where $x = (x_1, x_2, x_3) \in R^3$. Then f is clearly a linear functional on R^3 (here considered as 3-dimensional Euclidean space) and f is continuous, since

$$|f(x)| \leqslant \sqrt{3}\, \|x\| \quad \text{for all} \quad x \in R^3.$$

In fact $\|f\| = \sqrt{3}$, as is seen on putting $x = (1, 1, 1)$. By theorem 9, $\operatorname{Ker}(f)$ is a closed hyperplane in R^3 and in elementary geometry it is just the 'plane' $x_1 + x_2 + x_3 = 0$ through the origin.

In 3-space we often want to consider the distance of a plane (which is defined by a linear functional) from the origin. We may do this in a general normed space—finding in fact that there is a close connection with the norm of the bounded linear functional defining the hyperplane.

Theorem 10. *Let $f \not\equiv 0$ be in X^* and write*

$$M = \{x \mid f(x) = 1\}.$$

If d is the distance of the hyperplane M from the origin then $d = 1/\|f\|$.

Proof. By definition of distance we have $d = \inf\{\|x\| \mid f(x) = 1\}$. For every $x \in M$ we have $\|x\| \cdot \|f\| \geqslant 1$, $\|x\| \geqslant 1/\|f\|$ and so $d \geqslant 1/\|f\|$.

Now let $k > 1$ be arbitrary. From the definition of $\|f\|$ there exists $y \neq \theta$ such that

$$\frac{|f(y)|}{\|y\|} > \frac{\|f\|}{k}.$$

Hence $k/\|f\| > \|x\|$, where $x = y/f(y)$; $f(x) = 1$. Therefore $k/\|f\| > d$ and since k was arbitrary we have $1/\|f\| \geqslant d$. Combining our two inequalities we get $d = 1/\|f\|$.

Exercises 2

1. Let X, Y be normed spaces and $A \in L(X, Y)$. If A is continuous at a single point $x_0 \in X$ show that $A \in B(X, Y)$.
2. Show that $c_0^* \sim l_1$ and $l_p^* \sim l_q$, $1 < p < \infty$, $1/p + 1/q = 1$.
3. Let X be a p-normed space and Y a normed space. Suppose $A : X \to Y$ is linear on X. Prove that A is continuous if and only if there exists a constant M such that

$$\|A(x)\| \leqslant M \|x\|^{1/p} \quad \text{for all} \quad x \in X.$$

(Thus we may define $\|A\| = \sup\{\|A(x)\| . \|x\|^{-1/p} \mid x \neq \theta\}$ for such a continuous linear operator A.)
4. Let X be a normed space and M a closed subspace. Prove that the mapping $x \to E_x$ is a linear surjection, from X onto the factorspace X/M, which has norm 1.
5. Give examples of two norms, two seminorms and two linear functionals on l_∞.
6. Let C^n have the 'max norm', $\|x\| = \max\{|x_k| \mid 1 \leqslant k \leqslant n\}$, $n \geqslant 1$ being fixed. Suppose $a = (a_i)$ is fixed in C^n. Prove that

$$f(x) = \sum_{k=1}^{n} a_k x_k$$

defines an element of $(C^n)^*$. Prove that the image of the closed unit sphere in C^n is a closed disc in C^1.
7. Let $f : R \to R$ be continuous and additive, i.e. $f(x+y) = f(x) + f(y)$ for all $x, y \in R$. Prove that f must be of the form $f(x) = \lambda x$, for some $\lambda \in R$.
8. Let f, g be in $X^\dagger$. Prove that $\text{Ker}(f) = \text{Ker}(g)$ if and only if there exists $\lambda \neq 0$ such that $f(x) = \lambda g(x)$ on X.
9. Let X be normed and S a maximal closed subspace. Prove that there exists an $f \in X^*$ such that $\text{Ker}(f) = S$. As a hint take $x_0 \in X \sim S$ and show that $x = s + \lambda x_0$ for every $x \in X$, where $s \in S$ and λ is some scalar. Then we have a mapping $x \to \lambda$ and this has the required properties.
10. Let X be a finite dimensional normed space. Prove that $X^* = X^\dagger$. The following hints may be helpful.

   (i) $x \in X$ implies $x = \sum_{1}^{n} \lambda_k b_k$, where $\{b_1, ..., b_n\}$ is a Hamel base.

   (ii) $f \in X^\dagger$ implies $f(x) = \sum_{1}^{n} \lambda_k f(b_k)$, so $|f(x)| \leqslant M_n \left(\sum_{1}^{n} |\lambda_k|^2\right)^{\frac{1}{2}}$.

   (iii) $T(\lambda) = \left\|\sum_{1}^{n} \lambda_k b_k\right\|$ is a continuous function of $\lambda = (\lambda_1, ..., \lambda_n)$ on the compact set $K = \left\{\lambda \in C^n \mid \sum_{1}^{n} |\lambda_k|^2 = 1\right\}$. Hence T has a minimum $m \geqslant 0$.

   (iv) $m > 0$, otherwise $\lambda_1 b_1 + ... + \lambda_n b_n = \theta$ for some $\lambda \in K$.

   (v) $0 < m \leqslant T(\lambda)$ on K implies $\left(\sum_{1}^{n} |\lambda_k|^2\right)^{\frac{1}{2}} \leqslant m^{-1}\|x\|$. By (ii), $|f(x)| \leqslant m^{-1} M_n \|x\|$ on X, whence $f \in X^*$.

3. The Banach–Steinhaus theorem

The Banach–Steinhaus theorem (theorem 12 below) is one of the fundamental results in the theory of normed spaces. It has many and varied applications in several parts of analysis, some of which we shall be giving later. We first establish a result which is somewhat more general than theorem 12.

Theorem 11. *Let X be a second category p-normed space. Suppose F is a family $\{q\}$ of lower semicontinuous seminorms q such that*

$$q(x) \leqslant M(x) < \infty,$$

for each $x \in X$ and all $q \in F$.

Then there exists a constant M, independent of x and q such that

$$q(x) \leqslant M \|x\|^{1/p}$$

for all $x \in X$ and all $q \in F$.

Proof. In the theorem we are denoting the p-norm in X by $\|x\|$ so that $\|x\| = 0$ only if $x = \theta$, $\|\lambda x\| = |\lambda|^p \|x\|$, and $\|x+y\| \leqslant \|x\| + \|y\|$, where $p > 0$. Now with $d(x,y) = \|x-y\|$, (X,d) is a metric space, so by theorem 25, chapter 2, there exists a closed sphere $S[a,r]$ and a constant H such that $q(x) \leqslant H$ on $S[a,r]$ for all $q \in F$.

Take $x \in X$ such that $\|x\| > 0$. Then

$$q(sx) = q(sx+a-a) \leqslant q(sx+a) + q(a)$$

where $s = (r/\|x\|)^{1/p}$. Since $a \in S[a,r]$ we have $q(a) \leqslant H$ for all $q \in F$. Also, $sx + a \in S[a,r]$, since $\|sx\| = r$. Consequently, $q(sx) = sq(x) \leqslant 2H$ for all $q \in F$ and so, for such q,

$$q(x) \leqslant 2H \|x\|^{1/p} r^{-1/p}. \tag{8}$$

The case $x = \theta$ is also included in (8); both sides being zero. Our result follows, with $M = 2Hr^{-1/p}$.

Corollary. *Let X be as in the theorem and suppose (q_n) is a sequence of continuous seminorms such that there exists on X*

$$\lim_n q_n(x) = q(x), \quad say. \tag{9}$$

Then q is a continuous seminorm on X.

Proof. It is clear that q is a seminorm on X, for example

$$q_n(x+y) \leqslant q_n(x) + q_n(y) \quad \text{implies} \quad q(x+y) \leqslant q(x) + q(y),$$

on letting $n \to \infty$. By (9) the convergence of $(q_n(x))$ implies its boundedness and so by theorem 11 we have $q_n(x) \leqslant M \|x\|^{1/p}$ for all n and all $x \in X$. Hence $q(x) \leqslant M \|x\|^{1/p}$, so that q is continuous on X.

The following corollary to theorem 11 is what we shall refer to as the Banach–Steinhaus theorem.

Theorem 12 (Banach–Steinhaus). *If (A_n) is a sequence of bounded linear operators each defined on a Banach space X into a normed space Y and*

$$\limsup_n \|A_n(x)\| < \infty \quad \text{on } X, \tag{10}$$

then $\sup_n \|A_n\| < \infty$, i.e. the sequence $(\|A_n\|)$ of norms is bounded.

Proof. A Banach space is complete and normed and so a second category 1-normed space. For each n, $q_n(x) = \|A_n(x)\|$ defines a continuous seminorm q_n, since $A_n \in B(X, Y)$. Hence by (10) we may apply theorem 11 and obtain $q_n(x) \leqslant M \|x\|$ for all n and all $x \in X$. This implies $\|A_n\| \leqslant M$ for all n, as required.

Corollary. *Replace (10) of theorem 12 by the existence on X of*

$$\lim_n A_n(x) = A(x), \quad \text{say.}$$

Then A is a bounded linear operator on X into Y.

Proof. The proof follows exactly the same lines as that of the corollary to theorem 11, and so is omitted.

We remark that theorem 12 and its corollary will usually be applied for the special cases in which $Y = C$ or R, so that $\|A_n(x)\|$ is replaced by $|A_n(x)|$.

Let us give a simple example of the use of the last corollary.

Example 7. If for every $x \in l_1$ the series $\Sigma a_k x_k$ converges, then $a \in l_\infty$.

To show this we may define

$$f_n(x) = \sum_1^n a_k x_k,$$

which for each n is clearly a bounded linear functional on l_1. By hypothesis there exists $\lim_n f_n(x) = \Sigma a_k x_k = f(x)$, say, for every $x \in l_1$. The corollary yields $|f(x)| \leqslant M \|x\| = M \Sigma |x_k|$. Now put $x_n = \operatorname{sgn} a_n$, $x_k = 0$ for $k \neq n$. Then $f(x) = |a_n| \leqslant M$, $n = 1, 2, \ldots$, i.e. $a \in l_\infty$.

A completely elementary proof might run as follows: suppose $a \notin l_\infty$. Then a sequence $n_1 < n_2 < \ldots$ can be found such that $|a_{n_k}| > k^2$, $k = 1, 2, \ldots$. Put $x_{n_k} = 1/a_{n_k}$, $x_n = 0$ otherwise. Then $x \in l_1$ since $\|x\| < \pi^2/6$, but $\Sigma a_k x_k = 1 + 1 + \ldots$ diverges to ∞. This contradiction shows that we must have $a \in l_\infty$.

A pleasing example of the use of the Banach–Steinhaus theorem is now given. We assume that the reader has a slight knowledge of the basic idea of a Fourier series. Very little needs to be known—where necessary we refer the reader to the book by K. Knopp, *Theory and Application of Infinite Series* (Blackie, 1964).

Example 8. There exists a continuous function of period 2π whose Fourier series diverges at 0.

Let f be a continuous function of period 2π and X the Banach space of all such functions, with $\|f\| = \max\{|f(t)|\,|\,t \in [0, 2\pi]\}$. The Fourier coefficients of f are, by definition,

$$a_k = \frac{1}{\pi} \int_0^{2\pi} f(t) \cos kt \, dt, \quad b_k = \frac{1}{\pi} \int_0^{2\pi} f(t) \sin kt \, dt \quad (k = 0, 1, \ldots).$$

The Fourier series of f is the *formal* series

$$\frac{a_0}{2} + \sum_{k=1}^{\infty} (a_k \cos kx + b_k \sin kx). \tag{11}$$

It is formal in that it may not converge at all—if it does it may not converge to f.

When the nth partial sum of (11) is calculated (see Knopp, p. 359) at $x = 0$ it is found to be

$$s_n(f) = \frac{1}{2\pi} \int_0^{2\pi} D_n(t) f(t) \, dt, \tag{12}$$

where $\qquad D_n(t) = (\sin (n + \tfrac{1}{2}) t)/\sin \tfrac{1}{2} t.$

The function $D_n(t)$ is known as Dirichlet's kernel, after the great German mathematician P. G. Lejeune-Dirichlet (1805–59) who, in 1829, initiated rigorous investigations into the theory of Fourier series.

Now (12) defines, for each n, a continuous linear functional s_n on X. For s_n is obviously linear and it is easy to show that $|D_n(t)| \leqslant 2n + 1$. Also,

$$|s_n(f)| \leqslant \frac{\|f\|}{2\pi} \int_0^{2\pi} |D_n(t)| \, dt = \|f\| \, l_n, \quad \text{say.}$$

Hence $\|s_n\| \leqslant l_n$. We shall shortly show that $\|s_n\| = l_n$. Having done this we then argue as follows. Suppose $(s_n(f))$ converges for all $f \in X$. Then, by the Banach–Steinhaus theorem, we have that

$$\sup_n \|s_n\| = \sup_n l_n < \infty.$$

But it turns out that $l_n \to \infty$ as $n \to \infty$ (as we shall prove). This contradiction means that $(s_n(f))$ cannot converge for all $f \in X$, whence there exists an f such that $(s_n(f))$ diverges, which is our result.

We do not wish to spoil the beauty of the application with too much analytical detail, however necessary. Consequently we indicate the idea of the argument, referring the reader to the exercises for the refinements.

If we put $f_0 = \operatorname{sgn} D_n$ in (12) then we get $s_n(f_0) = l_n$. But this f_0 has a finite number of discontinuities so we replace it by a function f which is 'smoothed' at the discontinuities. In this way we obtain an $f \in X$ such that $\|f\| = 1$ and $s_n(f) > l_n - \epsilon$, for any $\epsilon > 0$. Hence $\|s_n\| = l_n$, since we know already that $\|s_n\| \leqslant l_n$.

Finally, we show $l_n \to \infty$ as $n \to \infty$. Let us write, for each n,

$$c_k = (4k+1)\pi/4n+2, \quad d_k = (4k+3)\pi/4n+2.$$

Then for $t \in [c_k, d_k]$ we have $\quad |D_n(t)| \geqslant \dfrac{\sqrt{2}}{t}$.

Using this we see that

$$l_n > \frac{1}{\pi\sqrt{2}} \sum_{k=0}^{2n} \int_{c_k}^{d_k} \frac{dt}{t} > \frac{\sqrt{2}}{\pi} \sum_{k=0}^{2n} \frac{1}{4k+3} \to \infty$$

as $n \to \infty$.

Applications of the Banach–Steinhaus theorem to matrix transformations will be given in chapter 7.

Exercises 3

1. Let X be a Banach space and q a lower semicontinuous seminorm on X. Prove that q is continuous on X.
2. If for every $x \in l_p$ $(1 < p < \infty)$ the series $\Sigma a_k x_k$ converges, prove that $a \in l_q$ $(1/p + 1/q = 1)$.
3. If $\Sigma a_k x_k$ converges, whenever $x \in c$, prove that $a \in l_1$.
4. If $\Sigma a_k x_k$ converges, whenever Σx_k converges, prove that $a \in BV$.
5. If $\Sigma |a_k x_k| < \infty$, whenever Σx_k converges, prove that $a \in l_1$.
6. Let $f_n \in X^\dagger$ and $|f_n(x)| \leqslant M\|x\|$ for all n and all x. Let

$$S = \{x \mid \lim_n f_n(x) \text{ exists}\}.$$

Prove that S is a closed subspace of X.
7. Prove the unproved assertions in example 8, section 3.

4. The open mapping and closed graph theorems

If a mapping $A : X \to Y$ is bijective then there exists $A^{-1} : Y \to X$ and A^{-1} is bijective. If, in addition, X and Y are linear spaces and A is linear, then it is easy to see that A^{-1} is also linear. Now assume that X, Y are Banach spaces and that A is bounded (i.e. continuous). Then we have the remarkable theorem of Banach which tells us that A^{-1} is also bounded. We shall deduce this theorem from the 'open mapping theorem', which asserts that a bounded linear surjective mapping between Banach spaces is always an open mapping. The important 'closed graph theorem' is then established as a consequence of Banach's theorem.

Theorem 13 (Open mapping theorem). *Let X, Y be Banach spaces and suppose $A \in B(X, Y)$ is surjective. Then A is an open mapping.*

Proof. Let G be open in X and let $y \in A(G)$ so that $y = A(x)$ for some $x \in G$. Now there exists $S(x, \delta) \subset G$, whence $A(S(x, \delta)) \subset A(G)$. *Provided* we can show that there is a sphere $S(y) \subset A(S(x, \delta))$ we will have $S(y) \subset A(G)$, so that $A(G)$ will be open. To ensure the provision it is enough to prove that there exists a sphere $S(\theta) \subset A(S_0)$, where S_0 is the unit sphere in X and $S(\theta)$ is a sphere about the origin in Y. For, having proved this, $A(S(x, \delta) - x)$ will contain a sphere about the origin, whence $A(S(x, \delta))$ will contain a sphere about y.

Now we proceed to prove that there exists $S(\theta) \subset A(S_0)$. Let

$$S_k = S(\theta, 2^{-k}) = \{x \mid \|x\| < 2^{-k}\} \quad (k = 0, 1, \ldots).$$

Now if $x \in X$ we may write $x = k(x/k)$, where $k = [2\|x\|] + 1$; the square bracket denoting the integer part of $2\|x\|$. Then $x/k \in S_1$ and so $x \in kS_1$. Hence

$$X = \bigcup_{k=1}^{\infty} kS_1.$$

Since A is onto, $A(X) = Y$, we get

$$Y = \bigcup_{k=1}^{\infty} kA(S_1).$$

But Y is complete and so second category, whence there exists k such that $kA(S_1)$ is not nowhere dense. Thus $\overline{A(S_1)}$ contains some sphere $S(a, r)$, say. We shall now show that

$$S(\theta, r) \subset \overline{A(S_0)}. \tag{13}$$

We have

$$S(\theta, r) \subset \overline{A(S_1)} - a \subset \overline{A(S_1)} - \overline{A(S_1)} \subset 2\overline{A(S_1)} = \overline{A(S_0)},$$

which proves (13). In the above chain of inclusions we have used the facts that $a \in \overline{A(S_1)}$, $\overline{A(S_1)}$ is convex and that $-y \in \overline{A(S_1)}$ whenever $y \in \overline{A(S_1)}$.

From (13) it now follows immediately that

$$S(\theta, r2^{-n}) \subset \overline{A(S_n)}. \tag{14}$$

Finally, we show that $S(\theta, r/2) \subset A(S_0)$. Take $y \in S(\theta, r/2)$. Then $y \in \overline{A(S_1)}$ by (14). Hence $\|y - y_1\| < r/4$ for some $y_1 \in A(S_1)$. Also, $y - y_1 \in S(\theta, r/4) \subset \overline{A(S_2)}$ and so $\|y - y_1 - y_2\| < r/8$ for some $y_2 \in A(S_2)$. Continuing, we find

$$\left\| y - \sum_{k=1}^{n} y_k \right\| < r/2^{n+1}, \tag{15}$$

where $y_k \in A(S_k)$, whence $y_k = A(x_k)$ for some $x_k \in S_k$. From (15) we get

$$y = \Sigma A(x_k),$$

and since $\|x_k\| < 2^{-k}$, $\Sigma \|x_k\| < 1$, we see that $\Sigma x_k = x$, say. Also $\|x\| \leqslant \Sigma \|x_k\| < 1$, $x \in S_0$, and by continuity of A,

$$Ax = \Sigma A(x_k) = y.$$

Thus $y \in S(\theta, r/2)$ implies $y = Ax$ for some $x \in S_0$, i.e. $y \in A(S_0)$, which proves that $S(\theta, r/2) \subset A(S_0)$. From our earlier remarks this inclusion is enough to allow us to deduce the open mapping theorem.

Theorem 14 (Banach's theorem). *Let X, Y be Banach spaces and suppose $A \in B(X, Y)$ is bijective. Then $A^{-1} \in B(Y, X)$.*

Proof. The result follows immediately from theorem 13, for A bijective, continuous and open implies A bicontinuous. Thus A is a linear homeomorphism.

Our next theorem has useful applications, though it is often a matter of personal taste as to whether to employ it rather than the Banach–Steinhaus theorem for some particular problem.

Before the theorem is given we recall that the graph of a mapping $A : X \to Y$ is $\{(x, Ax) | x \in X\}$, which is a subset of $X \times Y$. If X, Y are normed spaces we may turn $X \times Y$ into a normed space with

$$\|(x, y)\| = (\|x\|^2 + \|y\|^2)^{\frac{1}{2}},$$

defining $(x, y) + (x', y') = (x + x', y + y')$, $\lambda(x, y) = (\lambda x, \lambda y)$. It is clear that $(x_n, y_n) \to (x, y)$ if and only if $x_n \to x$ and $y_n \to y$. Hence if X, Y are Banach spaces then $X \times Y$ is also a Banach space.

When A is continuous it is easy to show that the graph of A is a closed subset of $X \times Y$. With extra conditions we have a more complete result:

Theorem 15 (Closed graph theorem). *Let X, Y be Banach spaces and let $A \in L(X, Y)$. Then A is continuous if and only if its graph is closed.*

Proof. (i) Let A be continuous and let $G(A)$ be the graph of A. We show that $G(A)$ is closed—actually the linearity of A is superfluous in this part of the theorem. Let $(x, y) \in \overline{G(A)}$. Then there exists $x_n \in X$ such that $x_n \to x$ and $Ax_n \to y$. But $Ax_n \to Ax$ and so $y = Ax$, whence $(x, y) = (x, Ax) \in G(A)$.

(ii) Suppose $G(A)$ is closed. Then $G(A)$ is a Banach space, being a closed subspace of the Banach space $X \times Y$. Now consider the mapping $f : G(A) \to X$ given by

$$f((x, Ax)) = x.$$

This mapping is clearly a linear injection. Also, f is continuous, since $\|f((x, Ax))\| = \|x\| \leqslant \|(x, Ax)\|$. Hence we may apply Banach's theorem (theorem 14) and infer that $f^{-1} : X \to G(A)$ is continuous. Finally,

$$\|Ax\| \leqslant \|(x, Ax)\| = \|f^{-1}(x)\| \leqslant \|f^{-1}\| \, \|x\|,$$

so that A is bounded and hence continuous.

Let us give a simple application of the closed graph theorem.

Example 9. Suppose (a_{nk}), $n, k = 1, 2, \ldots$ is a given infinite matrix of complex numbers a_{nk}. Let $y_n = \Sigma_k a_{nk} x_k$ converge for each n and let $y = (y_n) \in l_\infty$, for each $x \in l_\infty$. Then the operator A defined by $y = Ax$ is an element of $B(l_\infty, l_\infty)$.

First, the convergence of $\Sigma a_{nk} x_k$ for each n and each $x \in l_\infty$ implies $\Sigma |a_{nk}| < \infty$ for each n—just put $x_k = \operatorname{sgn} a_{nk}$, $k = 1, 2, \ldots$ for each n. Thus $A_n(x) = \Sigma a_{nk} x_k$ defines‡ an element $A_n \in l_\infty^*$. The operator A is obviously linear, so its continuity has to be established. By the closed graph theorem it is enough to show that $G(A)$ is closed. Take any $(x, y) \in \overline{G(A)}$. Then there exists $x^{(i)} \in l_\infty$ such that $x^{(i)} \to x$, $Ax^{(i)} \to y$ as $i \to \infty$. Now the continuity of A_n implies $A_n(x^{(i)}) \to A_n(x)$, $i \to \infty$, each

‡ This is also a consequence of the corollary to theorem 12, section 3, chapter 4.

n and $Ax^{(i)} \to y$ clearly implies $A_n(x^{(i)}) \to y_n$, $i \to \infty$, each n. Hence $y_n = A_n(x)$, $y = Ax$ and so $(x, y) \in G(A)$, which proves that $\overline{G(A)} \subset G(A)$, i.e. $G(A)$ is closed.

Exercises 4

1. Let $(X, \|.\|)$, $(X, \|.\|')$ be Banach spaces such that $\|x\| \leqslant K \|x\|'$ for all $x \in X$ and some constant K. Prove that there exists a constant M such that $\|x\|' \leqslant M \|x\|$ for all $x \in X$, and hence show that the two norms generate the same metric topology on X.
2. Let $A_1 \in B(X_1, X_3)$, $A_2 \in B(X_2, X_3)$, where X_1, X_2, X_3 are Banach spaces. If the equation $A_1(x) = A_2(y)$ has for every x a unique solution $y = A(x)$, prove that A is a bounded linear operator on X_1 into X_2 (use the closed graph theorem).
3. X, Y are Banach spaces and the operator $A_0 \in B(X, Y)$ is bijective. If the operator $A_1 \in B(X, Y)$ is such that $\|A_0^{-1}\| \|A_1\| < 1$, prove that the operator $A_0 + A_1$ has an inverse (using the contraction principle of chapter 2).

5. The Hahn–Banach extension theorem‡

One of the basic theorems of linear space theory is shortly to be proved. It is important in that it guarantees the existence of linear extensions to the whole space of linear functionals defined on a subspace. The proof depends in part on the use of Zorn's lemma or some equivalent axiom.

First we prove the theorem for real linear spaces—the extension to complex linear spaces is fairly straightforward.

Theorem 16 (Hahn–Banach). *Let X be a real linear space and M a subspace. Suppose p is subadditive on X and $p(\alpha x) = \alpha p(x)$, whenever $\alpha \geqslant 0$ and $x \in X$. If f is a linear functional on M and*

$$f(x) \leqslant p(x) \quad on\ M$$

then there exists a linear extension g of f to X such that

$$g(x) \leqslant p(x) \quad on\ X.$$

Proof. Suppose $M \neq X$, otherwise there is nothing to prove. Take $z \in \sim M$ and $x, y \in M$. Then $f(x) - f(y) \leqslant p(x - y) \leqslant p(x + z) + p(-y - z)$ so that $-p(-y - z) - f(y) \leqslant p(x + z) - f(x)$. Hence for each $x \in M$,

$$s = \sup_{y \in M} \{-p(-y - z) - f(y)\} \leqslant p(x + z) - f(x),$$

‡ Proved independently by Banach and H. Hahn (1879–1934), the eminent German mathematician.

whence $$s \leqslant \inf_{x \in M} \{p(x+z) - f(x)\}.$$

Thus there is a number t such that, for all $y \in M$,

$$-p(-y-z) - f(y) \leqslant t \leqslant p(y+z) - f(y). \tag{16}$$

Now define the subspace $M_z = \{x + \alpha z \,|\, x \in M, \, \alpha \text{ real}\}$. Then if $w \in M_z$ we have $w = x + \alpha z$ and the uniqueness of this representation is readily checked. Define the function h on M_z by

$$h(w) = f(x) + \alpha t.$$

Then h is obviously linear on M_z and so is a linear extension of f from M to M_z. We now observe that

$$h(w) \leqslant p(w) \quad \text{on} \, M_z.$$

This follows immediately from (16) on putting $y = x/\alpha$, $\alpha \neq 0$ and splitting the cases $\alpha > 0$ and $\alpha < 0$.

If it happens that $M_z = X$ then we finish; if not we may repeat the process of extension, but what guarantee is there that we shall ever extend to the whole space X? It is here that we need Zorn's lemma or one of its variants. We use Zorn's lemma (see chapter 1). Let P be the set of all ordered pairs (M', h'), where M' is a subspace containing M and h' is an extension of f to M' such that $h' \leqslant p$ on M'. Partially order P by defining

$$(M', h') \leqslant (M'', h'') \quad \text{if and only if} \quad M' \subset M'', h' = h'' \text{ on } M'.$$

Let $S = \{(M_\alpha, h_\alpha)\}$ be a totally ordered subset of P. Then it is easy to show that S has an upper bound $(\cup_\alpha M_\alpha, H)$, where

$$H(x) = h_\alpha(x) \quad \text{for} \quad x \in M_\alpha.$$

The point here is that $\cup M_\alpha$ is a subspace because of the total ordering of S.

By Zorn's lemma P has a maximal element $(\mathcal{M}, \mathcal{H})$, say. All we need now to complete the proof is to show that $\mathcal{M} = X$. Suppose that $\mathcal{M} \neq X$. Then, by the first part of our proof, we can *extend* in such a way that $(\mathcal{M}_z, \mathcal{H}_z) \geqslant (\mathcal{M}, \mathcal{H})$. Since $(\mathcal{M}, \mathcal{H})$ is maximal we must have $\mathcal{M}_z = \mathcal{M}$, which contradicts the fact that $\mathcal{M}_z \supset \mathcal{M}$, strictly. Hence the supposition that $\mathcal{M} \neq X$ is false, and the proof is completed on writing $g = \mathcal{H}$.

Corollary 1. *Every bounded linear functional defined on a subspace of a real normed linear space X can be extended linearly with preservation of the norm to the whole of X.*

Proof. We have $|f(x)| \leqslant \|f\| \|x\|$ on M, say. Hence there exists g such that $g(x) \leqslant \|f\| \|x\|$ on X. Putting $-x$ in place of x we deduce that $|g(x)| \leqslant \|f\| \|x\|$. Also, for $x \in M$ we have $g(x) = f(x)$, and so $\|g\| = \|f\|$.

Corollary 2. *If $X \neq \{\theta\}$ is a real normed linear space, then there are always non-trivial continuous linear functionals on X.*

Proof. In the Hahn–Banach theorem take $p(x) = \|x\|$. Now take any $x_0 \neq \theta$ and define $M = \{\alpha x_0 | \alpha \text{ real}\}$. Then M is a subspace and if we set $f(\alpha x_0) = \alpha \|x_0\|$ then f is linear and continuous on M. Hence we may extend f to the whole of X with preservation of the norm (corollary 1). Thus each non-zero element x_0 of X generates a continuous linear functional on X whose value at x_0 is $\|x_0\| > 0$.

Corollary 3. *Let X be a real normed linear space and suppose $f(x) = 0$ for all $f \in X^*$. Then we must have $x = \theta$.*

Proof. If $x \neq \theta$ there exists, by corollary 2, $f \in X^*$ such that $f(x) > 0$, whereas we are assuming $f(x) = 0$ for all $f \in X^*$. Hence the result.

Now we give the complex version of the Hahn–Banach theorem.

Theorem 17. *Let X be a complex linear space, M a subspace and let p be a seminorm on X. Suppose f is a complex-valued linear functional on M such that $|f(x)| \leqslant p(x)$ on the subspace M. Then there exists a linear extension g of f to X such that $|g(x)| \leqslant p(x)$ on X.*

Proof. Write $f = f_1 + if_2$, so that f_1, f_2 are real linear functionals on M, considered as a real linear space. Also, $f_1(ix) = -f_2(x)$ on M and $f_1(x) \leqslant |f(x)| \leqslant p(x)$ on M, so f_1 extends to g_1 with $g_1(x) \leqslant p(x)$ on X, using the real Hahn–Banach theorem. Now put $g(x) = g_1(x) - ig_1(ix)$ so that $g(x) = f(x)$ on M. Also, $g(ix) = ig(x)$, since g_1 is real linear, whence g is complex linear on X.

Finally, if $g(x) = 0$ then $|g(x)| \leqslant p(x)$—for p is non-negative. If $g(x) \neq 0$ let $\alpha = \arg g(x)$, so that

$$|g(x)| = g(x) e^{-i\alpha} = g(x e^{-i\alpha}) = g_1(x e^{-i\alpha})$$
$$\leqslant p(x e^{-i\alpha}) = p(x),$$

since $g(x e^{-i\alpha})$ is real and p is absolutely homogeneous.

It is now clear that the complex versions of the corollaries to theorem 16 are also valid.

As an application of the Hahn–Banach theorem we prove the Riesz‡ representation theorem for the elements of $C^*[0, 1]$.

Theorem 18 (Riesz representation). *If $f \in C^*[0, 1]$ then there exists a function g, of bounded variation on $[0, 1]$, such that*

$$f(x) = \int_0^1 x(t)\, dg(t) \quad \text{for all} \quad x \in C[0, 1],$$

$$\|f\| = V(g) = \text{total variation of } g \text{ on } [0, 1].$$

Proof. Before the proof we remark that we are assuming certain basic properties of Riemann–Stieltjes integrals $\int_0^1 x(t)\, dg(t)$. See, for example, Rudin (1964). Primarily, we need to know that if $x \in C[0, 1]$ and $g \in BV[0, 1]$ then $\int_0^1 x(t)\, dg(t)$ exists and

$$\lim_{\mu(P) \to 0} \sum_{i=1}^n x(c_i)\,(g(t_i) - g(t_{i-1})) = \int_0^1 x(t)\, dg(t),$$

where $P = \{0 = t_0, t_1, \ldots, t_n = 1\}$, $t_0 < t_1 < \ldots < t_n$, is a partition of $[0, 1]$; $\mu(P) = \max(t_i - t_{i-1})$, $1 \leqslant i \leqslant n$ and $c_i \in [t_{i-1}, t_i]$. It is to be understood that the limit exists for every partition P and every choice of c_i.

Now consider $C[0, 1]$ as a subspace of the Banach space M of bounded functions on $[0, 1]$ with $\|x\| = \sup\{|x(t)| \,|\, 0 \leqslant t \leqslant 1\}$, $x \in M$. If $f \in C^*[0, 1]$ then, by the Hahn–Banach theorem, there exists a linear extension F to M such that $f(x) = F(x)$ on $C[0, 1]$ and $\|F\| = \|f\|$.

Let χ_t denote the characteristic function of $[0, t]$, $0 \leqslant t \leqslant 1$, i.e. $\chi_t(y) = 1$ $(0 \leqslant y \leqslant t)$, $\chi_t(y) = 0$ $(t < y \leqslant 1)$. Then $\chi_t \in M$ for each $t \in [0, 1]$ and we write $F(\chi_t) = g(t)$. This function g is the function appearing in the theorem. Now write $\epsilon_i = \text{sgn}\,[g(t_i) - g(t_{i-1})]$. Then $|g(t_i) - g(t_{i-1})| = [g(t_i) - g(t_{i-1})]\,\epsilon_i$ and so

$$\sum_{i=1}^n |g(t_i) - g(t_{i-1})| \leqslant \|F\| \cdot \left\| \sum_{i=1}^n [\chi_{t_i} - \chi_{t_{i-1}}]\,\epsilon_i \right\|.$$

If $y \in [0, 1]$ then $y \in (t_{j-1}, t_j]$ for some j, whence

$$\sum_{i=1}^n [\chi_{t_i}(y) - \chi_{t_{i-1}}(y)]\,\epsilon_i = \epsilon_j.$$

---

‡ F. Riesz (1880–1956), the famous Hungarian mathematician, was one of the pioneers in functional analysis.

Consequently
$$\sum_{i=1}^{n} |g(t_i) - g(t_{i-1})| \leqslant \|F\|, \tag{17}$$

whence $g \in BV[0, 1]$.

Now define
$$z_n = \sum_{r=1}^{n} x(r/n)\,(\chi_{r/n} - \chi_{(r-1)/n}).$$

Then if $u \in [0, 1]$ we have $u \in ((r-1)/n, r/n)$ for some r and so
$$z_n(u) = x(r/n).$$

Since x is continuous and hence uniformly continuous on $[0, 1]$ it follows that $\|x - z_n\| \to 0$ $(n \to \infty)$. Consequently $F(z_n) \to F(x) = f(x)$ and so
$$f(x) = \lim_{n \to \infty} \sum_{r=1}^{n} x(r/n)\left(g(r/n) - g\left(\frac{r-1}{n}\right)\right)$$
$$= \int_0^1 x(t)\,dg(t). \tag{18}$$

From (18), we obtain
$$|f(x)| \leqslant \max |x(t)| . V(g) = \|x\|\, V(g),$$

so that $\|f\| \leqslant V(g)$, which, coupled with (17), yields $\|f\| = V(g)$.

In theorem 18 the function $g \in BV$ is not unique. For example we may add an arbitrary constant to g and still get the same representation for f. Thus we cannot identify $C^*[0, 1]$ with $BV[0, 1]$.

To obtain an identification of $C^*[0, 1]$ we shall need to use two well-known properties of functions g in $BV[0, 1]$. The first is that $g(t+0) = \lim_{x \to t+} g(x)$ exists for each $t \in [0, 1]$, and the second is that the set of discontinuities of g is at most countable. Now write
$$\widehat{BV}[0, 1] = \{G \in BV \mid G(0) = 0 \quad \text{and} \quad G(t+0) = G(t), 0 < t < 1\}.$$

Then $\widehat{BV}$ is a subspace of BV. Define a function G by $G(0) = 0$, $G(1) = g(1) - g(0)$ and $G(t) = g(t+0) - g(0)$, $0 < t < 1$. Then
$$G(t) = g(t) - g(0)$$

for $t = 0$ and $t = 1$ and for each $t \in (0, 1)$ at which g is continuous. Hence for every $x \in C[0, 1]$ we have
$$\int_0^1 x(t)\,dg(t) = \int_0^1 x(t)\,dG(t).$$

Now it is clear that $G \in \widehat{BV}$. Hence with each $f \in C^*[0, 1]$ is associated a $G \in \widehat{BV}$ such that
$$f(x) = \int_0^1 x(t)\,dG(t) \quad \text{for all} \quad x \in C[0, 1].$$

The function G is in fact unique. For suppose that $h \in \widehat{BV}$ is such that $f(x) = \int_0^1 x\,dh$, for all $x \in C[0,1]$. Putting $x \equiv 1$ we get $G(1) = h(1)$, since $G(0) = h(0) = 0$. Now take $0 < c < 1$ and write $H(t) = G(t) - h(t)$. Then $\int_0^1 x\,dH = 0$, for all $x \in C[0,1]$. Choose x to be equal to 1 on $[0,c]$, equal to 0 on $[c+h,1]$ and complete the definition by joining the points $(c,1)$, $(c+h,0)$ with a straight line. Then $x \in C[0,1]$ and, on integration by parts,

$$0 = H(c) + \int_0^{c+h} x(t)\,dH(t)$$

$$= H(c) - H(c) - \int_0^{c+h} x'(t)\,H(t)\,dt$$

$$= \frac{1}{h}\int_0^{c+h} H(t)\,dt \to H(c+0) \quad \text{as} \quad h \to 0+.$$

Hence, for $0 < c < 1$, we have $H(c+0) = 0$, i.e. $G(c+0) = h(c+0)$, i.e. $G(c) = h(c)$. We have thus shown that $G = h$ on $[0,1]$, so that G is unique. It is now a simple matter to show that $\|f\|$ is equal to the total variation of G on $[0,1]$, and this completes the identification of $C^*[0,1]$ with $\widehat{BV}[0,1]$.

Bidual of a normed space. Another use of the Hahn–Banach theorem is in connection with the second dual space (or bidual) of a normed space X. The bidual is denoted by $X^{**} = (X^*)^*$. We are going to show that X may be identified with a certain subspace $\tilde{X}$ of X^{**}. One then thinks of X as being naturally embedded in its bidual X^{**}.

Theorem 19. *Let X be a normed linear space. Then X is isometrically isomorphic to a subspace $\tilde{X}$ of the bidual X^{**}.*

Proof. Let us denote elements of X by x, as usual, and elements of X^* by x^*. Define a linear functional $\tilde{x}$ on X^* by

$$\tilde{x}(x^*) = x^*(x).$$

Thus, with each $x^* \in X^*$ we associate the value of x^* at the point $x \in X$. That $\tilde{x}$ is linear on X^* is a consequence of the way the linear operations are defined in X^*. Now $\tilde{x}$ is also continuous on X^*:

$$|\tilde{x}(x^*)| = |x^*(x)| \leqslant \|x^*\|\,\|x\|,$$

whence $\|\tilde{x}\| \leqslant \|x\|$. Hence the map $x \to \tilde{x}$ defines a continuous linear operator on X into X^{**}. Now let $\tilde{X}$ denote the image of X under the map $x \to \tilde{x}$. All we need to show then is that the map is an isometry. By corollary 2 to the Hahn–Banach theorem there exists $y^* \in X^*$ such that $y^*(x) = \|x\|$ and $\|y^*\| = 1$. Hence

$$\|\tilde{x}\| = \sup_{x^* \neq \theta} \frac{|\tilde{x}(x^*)|}{\|x^*\|} \geqslant |\tilde{x}(y^*)| = \|x\|,$$

so that, by the earlier result $\|\tilde{x}\| \leqslant \|x\|$, we now have $\|\tilde{x}\| = \|x\|$. This proves the theorem.

In general we have $\tilde{X} \subset X^{**}$. If equality occurs we have a *reflexive space*.

Analytic vector-valued functions.

There is another use of the Hahn–Banach theorem which we should like to mention. This concerns a generalization of the familiar Liouville's theorem of complex variable, which asserts that a bounded integral function must be constant.

First, we define an analytic vector-valued function $x = x(z)$ on a domain D in the complex plane C. Let X be a Banach space and $x : D \to X$ a mapping of D into X. Then x is called analytic on D if and only if there exists

$$\lim_{h \to 0} \frac{x(z+h) - x(z)}{h} \quad (= x'(z))$$

for every $z \in D$ and $z + h \in D$. The limit is taken in the sense of the norm in X.

If $f \in X^*$ and x is analytic on D it is easy to see that fx defined by $(fx)(z) = f(x(z))$ is analytic on D in the usual sense of complex variable.

In the usual way we say that x is an *integral* function if it is analytic on C and we define it to be *bounded* if $\|x(z)\| \leqslant M$ for all $z \in C$. Now we prove the generalized Liouville theorem.

Theorem 20. *If $x : C \to X$, where X is a Banach space, and x is a bounded integral function, then x is a constant.*

Proof. For any $f \in X^*$ we have $|f(x(z))| \leqslant \|f\| \|x(z)\| \leqslant \|f\| M$ on C, so that fx is bounded. Since fx is also an integral function in the usual complex variable sense, the ordinary Liouville's theorem yields $f(x(z)) = f(x(z'))$ for any $z, z' \in C$, so $f(x(z) - x(z')) = 0$ for any $z, z' \in C$. By corollary 3 to the Hahn–Banach theorem it follows that $x(z) - x(z') = \theta$, $x(z) = x(z')$ for any $z, z' \in C$, i.e. x is a constant.

Exercises 5

1. M is a subspace of the normed space X and $y \in X$ is such that $d(y, M) \geqslant d > 0$. Use the Hahn–Banach theorem to prove that there exists an $f \in X^*$ such that $\|f\| \leqslant 1$, $f(y) = d$ and $f(m) = 0$ for all $m \in M$. As a hint, consider the subspace generated by M and y.

2. (Generalized limit in l_∞.) Consider the real linear space l_∞ and put $p(x) = \inf \left(\limsup_n k^{-1} \Sigma \{ x_{n+i_j} | 1 \leqslant j \leqslant k \} \right)$, where $x \in l_\infty$ and the inf is taken over all finite sets $i_1, i_2, \ldots, i_k$ of positive integers. Prove p is well-defined on l_∞, $p(\alpha x) = \alpha p(x)$, $\alpha \geqslant 0$, $x \in l_\infty$ and p is subadditive. Use the Hahn–Banach theorem to obtain a linear functional f on l_∞ with the following properties: $f(x) \geqslant 0$ whenever $x_n \geqslant 0$ for all n;

$$f((x_2, x_3, \ldots)) = f((x_1, x_2, \ldots));$$

$f(x) = 1$ if $x_n = 1$ for all n. Thus, on writing $f(x) = \operatorname{Lim} x_n$, we see that f has the properties of the ordinary 'lim' on c. The notation Lim indicates that the limit is in l_∞ not c.

3. Let X be a normed space and $x : D \to X$, where D is a domain in C. Prove that, if x is analytic on D and $f \in X^*$, then fx is analytic on D.

4. Let L be a simple closed path in C and let X be a Banach space. If $x : L \to X$ is continuous on L prove that $\displaystyle\int_L x(z)\, dz$ exists. The integral is defined as in complex variable theory except that the limit is taken in the norm. The completeness of X is used in the proof.

5. X is a Banach space and L is a simple closed path in C. If x is analytic within and continuous on L prove the generalized Cauchy theorem: $\displaystyle\int_L x(z)\, dz = \theta$. As a hint, use the ordinary Cauchy theorem and the Hahn–Banach theorem.

6. Weak convergence

Let (x_n) be a sequence in a normed space X. Then (x_n) is said to converge weakly to $x \in X$, written $x_n \to x$ (weakly), if and only if

$$f(x_n) \to f(x) \quad \text{as} \quad n \to \infty, \quad \text{for every } f \in X^*.$$

We call x the weak limit of (x_n).

If $x_n \to x$ in norm, i.e. $\|x_n - x\| \to 0$, then, since

$$|f(x_n) - f(x)| \leqslant \|f\| \, \|x_n - x\|$$

for every $f \in X^*$, we have $x_n \to x$ (weakly). Thus convergence in norm implies weak convergence to the same limit, and for this reason convergence in norm is often called strong convergence.

Theorem 21. (i) *The weak limit of a weakly convergent sequence is unique.*

(ii) *In general, weak convergence is not equivalent to strong convergence.*

(iii) *If $x_n \to x$ (weakly) then $\sup_n \|x_n\| < \infty$.*

Proof. (i) Suppose $x_n \to x$ (weakly) and $x_n \to y$ (weakly). Then $f(x-y) = f(x-x_n) + f(x_n-y) \to 0$, whence $f(x-y) = 0$, for every $f \in X^*$. By corollary 3 to the Hahn–Banach theorem (section 5, chapter 4) it follows that $x = y$.

(ii) Consider the space l_p $(1 < p < \infty)$. If we consult the table of dual spaces in section 2 of this chapter we find that every $f \in l_p^*$ may be written as $f(x) = \Sigma a_k x_k$ for all $x \in l_p$, where $a \in l_q$ and $1/p + 1/q = 1$. If we write $e_1 = (1, 0, 0, \ldots)$, $e_2 = (0, 1, 0, \ldots)$, $\ldots$, then, with the l_p norm we have $\|e_n - e_m\| = 2^{1/p}$ $(n \ne m)$, whence (e_n) is not strongly convergent. However, (e_n) is weakly convergent to θ. For, whenever $f \in l_p^*$ we have $f(e_n) = a_n$, and $a \in l_q$ implies $a_n \to 0$.

In the case $p = 1$ it happens that strong and weak convergence coincide. This will be proved as a corollary to a theorem of Schur on matrix transformations (see theorem 6, section 1, chapter 7).

(iii) $x_n \to x$ (weakly) implies $f(x_n - x) \to 0$ $(n \to \infty)$ for every $f \in X^*$. By corollary 2 to the Hahn–Banach theorem there exists $f_n \in X^*$ such that $f_n(x_n - x) = \|x_n - x\|$ and $\|f_n\| = 1$. For each $f \in X^*$ define $F_n(f) = f(x_n - x)$. Then (F_n) is a sequence of continuous linear functionals on the Banach space X^*. Since $f(x_n - x) \to 0$ it follows that $\limsup_n |F_n(f)| < \infty$ on X^*. The Banach–Steinhaus theorem (theorem 12, section 5, chapter 4) now yields $M = \sup_n \|F_n\| < \infty$. Hence

$$\|x_n - x\| = |f_n(x_n - x)| = |F_n(f_n)| \le \|F_n\| \|f_n\| \le M,$$

so that $\|x_n\| \le M + \|x\|$ for every n. This proves the theorem.

An important type of convergence can be defined in the dual space of a normed linear space X by saying that $f_n \to f$ (weak*) if and only if

(a) $f_n(x) \to f(x)$ as $n \to \infty$, for every $x \in X$,

and (b) $\sup_n \|f_n\| < \infty$.

This convergence is called weak* (weak star) convergence in X^*. If X is a Banach space, then (b) is superfluous, since the Banach–Steinhaus theorem shows that (b) is implied by (a).

Theorem 22. *Let X be a normed space which contains a countable subset $\{x_k\}$ whose linear hull is dense in X. Then*

$$d(f_1, f_2) = \sum_{k=1}^{\infty} \frac{|f_1(x_k) - f_2(x_k)|}{2^k(\|x_k\| + 1)}$$

is a metric on X^* and weak* convergence in X^* implies convergence in this metric.

Also, if S is the unit sphere $\{f \in X^* |\ \|f\| \leqslant 1\}$ then S is compact in the metric topology generated by d.

Proof. d is well-defined, since $d(f_1, f_2) \leqslant \|f_1 - f_2\| < \infty$ for any f_1, f_2 in X^*. If $d(f_1, f_2) = 0$ then $(f_1 - f_2)(x_n) = 0$ for $n = 1, 2, \ldots$. Now take any $x \in X$. Since l. hull $\{x_k\}$ is dense in X, there exists $y_{n_k} \to x\ (k \to \infty)$, where y_{n_k} is a linear combination of elements of $\{x_k\}$. Hence, by continuity of $f_1 - f_2$ we get $(f_1 - f_2)(x) = \lim_k (f_1 - f_2)(y_{n_k}) = 0$, so that $f_1 = f_2$. The other axioms for a metric space are trivially satisfied, whence d is a metric on X^*.

Now suppose $f_n \to f$ (weak*). Then $f_n(x) \to f(x)$ as $n \to \infty$, for every $x \in X$ and $M = \sup_n \|f_n\| < \infty$. Hence, for any N,

$$d(f_n, f) \leqslant \sum_{k=1}^{N} \frac{|f_n(x_k) - f(x_k)|}{2^k(\|x_k\| + 1)} + \sum_{k=N+1}^{\infty} \frac{M}{2^{k-1}}$$

$$= \Sigma_1 + \Sigma_2, \quad \text{say.}$$

Let $\epsilon > 0$ be given and choose N so large that $\Sigma_2 < \epsilon/2$. Then choose n_0 so large that $\Sigma_1 < \epsilon/2$ for $n > n_0$. Thus $f_n \to f$ in the metric d.

Finally, it is enough to show that the unit sphere S is sequentially compact, i.e. that every sequence (f_n) in S has a weak* convergent subsequence whose limit belongs to S. Now $(f_n(x_1))$ is bounded in C since $|f_n(x_1)| \leqslant \|f_n\| \|x_1\|$ for all n. Hence there is a convergent subsequence $(f_{n1}(x_1))$. Similarly $(f_{n1}(x_2))$ has a convergent subsequence $(f_{n2}(x_2))$. Continuing, we find a subsequence (f_{nn}) of (f_n) such that $(f_{nn}(x_k))$ converges in C for each k. Then $(f_{nn}(x_k))$ is Cauchy for each k, and since $\{x_k\}$ is dense in X it follows that $(f_{nn}(x))$ is Cauchy for each $x \in X$. Thus $f_{nn}(x) \to f(x)$, say, for each $x \in X$. It is obvious that f is linear and $\|f_{nn}\| \leqslant 1$ for all n implies $|f(x)| = \lim_n |f_{nn}(x)| \leqslant \|x\|$. Hence $\|f\| \leqslant 1$ and so $f \in S$, which proves the theorem.

We remark that what we have shown in theorem 22 is that if X is a certain type of normed space then there exists a metric d on X^* which makes S a compact subset of (X^*, d). Weak star convergence was used merely as a device to prove this result. A much more satisfactory result is obtained if the problem is approached topologically. It is possible to define, for any normed space X, a topology T^* (called the weak star topology) on X^* such that $S = \{f \in X^* | \|f\| \leqslant 1\}$ is a compact subset of the topological space (X^*, T^*). This general result is known as Alaoglu's theorem. Since the proof of this theorem requires rather more topology than we have at our disposal we refer the reader to more advanced works for the details.

Exercises 6

1. Prove that weak convergence is equivalent to strong convergence in the normed space C^n.

2. Suppose that the set $E \subset X^*$ is such that l. hull (E) is dense in X^*. Let (x_n) be a sequence in X such that $\sup_n \|x_n\| < \infty$ and $f(x_n) \to f(x)$ for every $f \in E$. Prove that $x_n \to x$ (weakly).

3. Let $x^{(n)} = (x_k^{(n)}) \in c$, $n = 1, 2, \ldots$. Prove that $x^{(n)} \to x = (x_k)$ (weakly) in c if and only if $\sup_n \|x^{(n)}\| < \infty$, $\lim_n x_k^{(n)} = x_k$, each k and

$$\lim_n (\lim_k x_k^{(n)}) = \lim_k x_k.$$

5

BANACH ALGEBRAS

1. Algebras and Banach algebras

This chapter presents a few basic concepts from the theory of Banach algebras. Our treatment is extremely limited in scope and is intended merely as a simple introduction. First we make some definitions. In all that follows our linear spaces are complex, unless specifically designated real.

Algebra

> *An algebra X is a linear space together with an internal operation of multiplication of elements of X, such that $xy \in X$, $x(yz) = (xy)z$, $x(y+z) = xy + xz$, $(x+y)z = xz + yz$, and $\lambda(xy) = (\lambda x)y = x(\lambda y)$, for scalar λ.*

In some algebras there exists a non-zero element e such that $ex = xe = x$ for all x. If such an e exists it is obviously unique and is then called the *identity* of the algebra.

An algebra is called commutative if it has the property that $xy = yx$, for all x, y.

Normed algebra

> *A normed algebra is an algebra which is normed, as a linear space, and in which $\|xy\| \leqslant \|x\| \cdot \|y\|$ for all x, y.*

The property $\|xy\| \leqslant \|x\| \cdot \|y\|$ is called the submultiplicative property of the norm.

Banach algebra

> *A Banach algebra is a complete normed algebra, i.e. an algebra which is a Banach space.*

Example 1. (i) C, with the usual addition and multiplication, and $\|x\| = |x|$, $x \in C$, is a commutative Banach algebra with identity. Here of course the norm is multiplicative, i.e. $\|xy\| = \|x\| \cdot \|y\|$.

(ii) R, with its usual structure, is a real commutative Banach algebra with identity.

Example 2. R^2, with co-ordinatewise linear operations and $\|x\| = (x_1^2 + x_2^2)^{\frac{1}{2}}$, $x = (x_1, x_2)$, is a real Banach space. R^2 becomes a real commutative Banach algebra with identity under 'complex' multiplication:

$$xy = (x_1 y_1 - x_2 y_2, x_1 y_2 + x_2 y_1),$$

where $x = (x_1, x_2)$, $y = (y_1, y_2)$. Note also that $\|xy\| = \|x\| \cdot \|y\|$.

Example 3. $C[0, 1]$ is an example of what is called a function algebra. We know already that $C[0, 1]$ is a Banach space. If we now define xy by $(xy)(t) = x(t) \cdot y(t)$, for $x, y \in C[0, 1]$, then, since the product of two continuous functions is continuous, we have $xy \in C[0, 1]$. It is now obvious that $C[0, 1]$ is a commutative algebra with identity. Also, $\|xy\| = \max |x(t) y(t)| = |x(t_0) y(t_0)| \leqslant \|x\| \cdot \|y\|$, whence $C[0, 1]$ is a Banach algebra.

Example 4. The space A of example 9, chapter 2, is an important Banach algebra called the *disc* algebra. Defining pointwise multiplication as in example 3 above we see that A is a commutative algebra, with identity: $f(z) = 1$ for $|z| \leqslant 1$. Now $\|f\| = \max |f(z)|$, the max being over $|z| \leqslant 1$, is a norm on A. Clearly the norm is submultiplicative. It remains to show that A is complete.

Let (f_n) be a Cauchy sequence in A. Then $(f_n(z))$ is Cauchy in C, for each $|z| \leqslant 1$. Hence $f_n(z) \to f(z)$ as $n \to \infty$, uniformly on $|z| \leqslant 1$ and so f is continuous on $|z| \leqslant 1$. Now let Γ be a simple closed path lying in $|z| < 1$. Since $f_n(z) \to f(z)$, uniformly on Γ, we have

$$\lim_n \int_\Gamma f_n(z)\, dz = \int_\Gamma f(z)\, dz.$$

By Cauchy's theorem, each integral on the left is zero, whence the limit is zero. Thus the integral on the right is zero for each Γ lying in $|z| < 1$. It follows by Morera's theorem that f is analytic on $|z| < 1$. Hence $f \in A$, so that A is complete.

Example 5. Let M^n denote the set of all $n \times n$ matrices $A = (a_{ij})$, with complex entries a_{ij}. Then M^n becomes a linear space with the definitions

$$A + B = (a_{ij} + b_{ij}), \quad \lambda A = (\lambda a_{ij}).$$

The usual definition of matrix multiplication:

$$(AB)_{ij} = \sum_{k=1}^{n} a_{ik} a_{kj},$$

makes M^n into an algebra. The matrix $I = (\delta_{ij})$, $\delta_{ii} = 1$, $\delta_{ij} = 0$ $(i \neq j)$, is the identity. As is well-known, matrix multiplication is not commutative. One may turn M^n into a normed algebra by defining

$$\|A\| = \max \left\{ \sum_{j=1}^{n} |a_{ij}| \,\big|\, 1 \leqslant i \leqslant n \right\}.$$

It is a simple matter to show that M^n is complete under this norm.

Example 6. Let X be a normed space. Then $B(X, X)$, the set of all bounded linear operators on X into itself, is a normed linear space. This result is a special case of theorem 7, section 2, chapter 4. Now define 'multiplication' of $A_1, A_2 \in B(X, X)$ as composition of operators: $(A_1 A_2)(x) = A_1(A_2 x)$. It is obvious that $A_1 A_2$ is linear whenever A_1 and A_2 are linear. Also, when A_1, A_2 are bounded,

$$\|(A_1 A_2)(x)\| \leqslant \|A_1\| \cdot \|A_2 x\| \leqslant \|A_1\| \|A_2\| \|x\|$$

and so $\|A_1 A_2\| \leqslant \|A_1\| \cdot \|A_2\|$. Since E, defined by $E(x) = x$ for all $x \in X$, is in $B(X, X)$ we see that $B(X, X)$ is a normed algebra with identity.

If X happens to be a Banach space then by theorem 7 (ii), section 2, chapter 4, we conclude that $B(X, X)$ is a Banach algebra with identity. $B(X, X)$ is an example of what is called an operator algebra.

In any algebra with identity e, an element which has an inverse is called a *unit*, i.e. x is a unit if and only if there exists an inverse y such that $xy = yx = e$. We write $y = x^{-1}$ and observe that x^{-1} is unique when it exists. Obviously e is a unit and θ is not a unit.

Consider example 1 (i) above. Every non-zero element of C is a unit. Similarly, every non-zero element of R^2 (in example 2) is a unit. However, in example 3, there are non-zero elements of $C[0, 1]$ which are not units, e.g. x given by $x(t) = 0$ on $[0, \frac{1}{2}]$, $x(t) = t - \frac{1}{2}$ on $[\frac{1}{2}, 1]$. These observations lead us to make the following definition.

Division algebra

A division algebra is an algebra with identity, in which every non-zero element is a unit.

C, and R^2 of example 2 are division algebras, both commutative. A four-dimensional non-commutative real division algebra was discovered by the celebrated Irish mathematician W. R. Hamilton

(1805–65). It was proved by Frobenius in 1878 that Hamilton's algebra (called the quaternions) was essentially the only finite dimensional non-commutative real division algebra.

Example 7 (The Quaternions). This algebra H (for Hamilton) is the real linear space R^4 provided with a special kind of multiplication which makes R^4 into a non-commutative division algebra. Following Hamilton we denote the unit vectors in R^4 by $1, i, j, k$. Precisely, we write $1 = (1, 0, 0, 0)$, $i = (0, 1, 0, 0)$, $j = (0, 0, 1, 0)$, $k = (0, 0, 0, 1)$. Once we have defined the products ij, jk, etc., we may multiply

$$xy = (x_1 + \ldots + kx_4)(y_1 + \ldots + ky_4)$$

in the usual way. The definition of multiplication of unit vectors is this:

$$1 \cdot m = m \cdot 1 = m \quad \text{for} \quad m = 1, i, j, k,$$

$$m^2 = -1 \quad \text{for} \quad m = i, j, k,$$

$$ij = -ji = k, \quad jk = -kj = i, \quad ki = -ik = j.$$

We note immediately that Hamilton has sacrificed commutative multiplication—a bold step in those far-off days. It is now trivial to check that H is an algebra with identity 1.

When operating with complex numbers $x = x_1 + ix_2$ it is useful to use the conjugate $\bar{x} = x_1 - ix_2$. Then $x\bar{x} = |x|^2 = x_1^2 + x_2^2$. Similarly, with each $x \in H$ we associate $\bar{x} = x_1 - ix_2 - jx_3 - kx_4$. Then

$$x\bar{x} = x_1^2 + x_2^2 + x_3^2 + x_4^2,$$

which yields the natural norm

$$\|x\| = (x\bar{x})^{\frac{1}{2}}.$$

Obviously H is complete under this norm. For any $x, y \in H$ one readily checks that $\overline{xy} = \bar{y} \cdot \bar{x}$, from which it follows that $\|xy\| = \|x\| \cdot \|y\|$. Now if $x \neq \theta$, then $\|x\| > 0$ and so $x\bar{x}\|x\|^{-2} = 1$, whence the inverse of x is $\bar{x}\|x\|^{-2}$. Collecting our results we see that H is a real non-commutative Banach division algebra.

Exercises 1

1. Suppose that a normed algebra has identity e. Prove that $\|e\| \geqslant 1$.
2. In a normed algebra, prove that multiplication is continuous, i.e. $x \to x_0$, $y \to y_0$ imply $xy \to x_0 y_0$, convergence being in the norm.
3. Let X be an algebra without identity. Consider the Cartesian product $Y = C \times X$. Define linear operations in Y in the usual co-ordinatewise fashion. Define multiplication in Y as follows:

$$(\lambda_1, x_1)(\lambda_2, x_2) = (\lambda_1 \lambda_2, \lambda_1 x_2 + \lambda_2 x_1 + x_1 x_2).$$

Prove that Y is an algebra with identity. Show also that X is iso-morphically embedded in Y, in the sense that X is isomorphic to a subset of Y. (See section 2 for the idea of isomorphic algebras.)

4. If X of question 3 above is also a Banach algebra, show how to define a norm in Y so that Y becomes a Banach algebra.

5. Let $l_1(Z) = \{x : Z \to C \mid \Sigma |x(k)| < \infty\}$, where in this question sums run from $k = -\infty$ to $k = +\infty$. Prove that $l_1(Z)$ is a linear space and that $\|x\| = \Sigma |x(k)|$ is a norm on $l_1(Z)$. Show also that xy defined by

$$(xy)(n) = \Sigma x(n-k)\,y(k)$$

makes $l_1(Z)$ into a Banach algebra. The multiplication just defined is usually called *convolution* and is frequently denoted by $x * y$, rather than xy.

6. Is A, of example 4, a division algebra?

7. Show that, for $n > 1$, M^n of example 5 is not a division algebra.

8. (i) If $x, y \in H$, the algebra of quaternions, prove that $\overline{xy} = \overline{y}\,\overline{x}$, and deduce that the norm in H is multiplicative:

$$\|xy\| = \|x\| \cdot \|y\|.$$

(ii) Let $x \neq \theta$ be in H. By solving the four equations obtained by equating coefficients in $xy = (1, 0, 0, 0)$, find the inverse y of x.

(iii) Show that there are infinitely many $x \in H$ such that $x^2 = -1$.

(iv) Prove that $G = \{1, -1, i, -i, j, -j, k, -k\}$ is a group under quaternion multiplication.

9. Let X be a commutative algebra with identity e. Suppose that x is not a unit and define

$$S = \{xy \mid y \in X\}.$$

Prove that

(i) $S \neq \varnothing$,

(ii) $e \notin S$,

(iii) S is a linear subspace of X,

(iv) $z \in X$, $s \in S$ imply $zs \in S$.

2. Homomorphisms and isomorphisms

Mappings between algebras which preserve the linear and multiplica-tive operations are of special importance. Such mappings are called

Homomorphisms

> Let X, Y be algebras. Then a map $f : X \to Y$ is called a homo-morphism if and only if $f(\lambda x + \mu y) = \lambda f(x) + \mu f(y)$ and $f(xy) = f(x) \cdot f(y)$, i.e. f is linear and multiplicative.

An isomorphism between algebras is defined to be a bijective homomorphism. A homomorphism $f : X \to C$ is usually called a scalar homomorphism. When we speak of a bounded homomorphism f on a normed algebra X we mean that $\|f(x)\|_Y \leqslant M\|x\|_X$ for some constant M and all $x \in X$. Usually we write this inequality as $\|f(x)\| \leqslant M\|x\|$, although the norms on each side of $\leqslant$ may be different.

It is an especially curious feature of the theory of Banach algebras that quite 'ordinary' hypotheses have 'extraordinary' conclusions. This remark is well-illustrated by theorem 1 below. The reason for such behaviour would seem to be that there is so much structure (both analytic and algebraic) in the hypotheses that 'ordinary' is perhaps not quite the right word to describe them.

We preface the proof of theorem 1 with a simple lemma.

Lemma. *Let X be a Banach algebra and suppose x is such that $\|x\| < 1$. Then there exists $y \in X$ such that $xy = x + y$.*

Proof. Since $\|x\| < 1$ and $\|x^k\| \leqslant \|x\|^k$, the series $-x - x^2 - x^3 - \ldots$ is absolutely convergent. But X is a Banach space and so the series converges (theorem 2, section 1, chapter 4). Let the sum of the series be y. Then
$$xy = -x^2 - x^3 - x^4 - \ldots = x + y.$$

Theorem 1. *Let X be a Banach algebra and $f : X \to C$ a scalar homomorphism. Then $|f(x)| \leqslant \|x\|$ for all $x \in X$. Thus a scalar homomorphism on a Banach algebra is necessarily a continuous functional.*

Proof. Suppose there exists $z \in X$ such that $|f(z)| > \|z\|$. Then $z \neq \theta$ and we may write $x = z/f(z)$, so that $f(x) = 1$, $\|x\| < 1$. By the lemma just proved above there exists y such that $xy = x + y$, whence $f(x).f(y) = f(x) + f(y)$, i.e. $f(y) = 1 + f(y)$, which gives a contradiction. Hence we must have $|f(x)| \leqslant \|x\|$ for all x.

It was remarked in section 1 of the present chapter that Frobenius had proved that the quaternion algebra H was the only finite dimensional non-commutative real division algebra, in the sense that every such algebra is isomorphic to H. Actually, Frobenius proved more, viz. that there are only three finite dimensional real division algebras: R, R^2 with complex multiplication, and H. The proof of this general result is rather intricate and we do not wish to go into the details of it. However, the one-dimensional case is quite simple and illustrates the idea.

Theorem 2. *Let X be a one-dimensional real division algebra. Then X is isomorphic to the algebra R.*

Proof. Let e be the identity in X and suppose that $\{b\}$ is a Hamel base for X. Since b^2 is in X, there exists a unique real λ such that $b^2 = \lambda b$, which implies $b(b - \lambda e) = \theta$. But $b \neq \theta$ and so there exists b^{-1}. Thus $(b^{-1}b)(b - \lambda e) = b^{-1}\theta = \theta$, whence $b = \lambda e$. Now each $x \in X$ is uniquely expressible in the form $x = \mu b = \mu \lambda e$. It is clear that the map $x \to \mu \lambda$ from X to R is linear, multiplicative and bijective, i.e. is an isomorphism.

In chapter 3, it was shown in theorem 3 that every n-dimensional linear space was isomorphic to C^n. It is natural to ask for some analogue of this result for n-dimensional algebras (the dimension of an algebra being its dimension as a linear space). This analogue is provided by the next theorem.

Theorem 3. *Every n-dimensional algebra X with identity e is isomorphic to an algebra of $n \times n$ matrices.*

Proof. Let $x \in X$. Associate with x the map $x^* : X \to X$, defined by $x^*(y) = xy$, for all $y \in X$. We may now construct a map $f : X \to L$, where $f(x) = x^*$ and L is the set of all the maps x^*. It is easy to check that f is linear on X. Let us show that f is multiplicative. Write $x = x_1 x_2$, so that for any y, we have

$$x^*(y) = xy = x_1(x_2 y) = x_1^*(x_2 y) = x_1^*(x_2^*(y)) = (x_1^* x_2^*)(y),$$

by definition of multiplication (composition) of maps. Hence $x^*(y) = (x_1^* x_2^*)(y)$ for all y, i.e. $x^* = x_1^* x_2^*$, i.e. $f(x_1 x_2) = f(x_1) . f(x_2)$.

Now f is injective, for $f(x_1) = f(x_2)$ implies $x_1^*(y) = x_2^*(y)$ for all y. Putting $y = e$, we get $x_1 = x_2$. By definition of L we also have $f(X) = L$, so that X is isomorphic to a certain algebra of maps of X into itself. Observe that we have not yet used the fact that X is n-dimensional. We complete the proof of the theorem as follows. Let $\{b_1, ..., b_n\}$ be a Hamel base for X. Then each y in X is of the form $y = \Sigma \lambda_i b_i$, the sum being over $1 \leqslant i \leqslant n$. Thus, for each x in X,

$$x^*(y) = \sum_i \lambda_i x^*(b_i) = \sum_i \lambda_i \sum_j a_{ij} b_j$$
$$= \sum_j b_j \left(\sum_i a_{ij} \lambda_i \right).$$

It is now easy to check that the map taking x to the matrix $A = (a_{ij})$ is an isomorphism.

Exercises 2

1. Let X be a finite dimensional real division algebra with identity e. Prove that every $x \in X$ satisfies a quadratic equation $\lambda_0 e + \lambda_1 x + \lambda_2 x^2 = \theta$, with real coefficients.

2. Let X be as in question 1, and also commutative. Show that to every $x \in X$ there correspond $\lambda, \mu \in C$ such that $x^2 - (\lambda + \mu)x + \lambda\mu e = \theta$. Deduce that x is of the form $x = \alpha e$, for some $\alpha \in C$.

3. Let $\|x\| \leqslant c < 1$, where x is an element of a Banach algebra with identity e. Prove that

   (i) $(e+x)^{-1} = \sum\limits_{n=0}^{\infty} (-1)^n x^n = e - x + x^2 - \ldots,$

   (ii) $\|(e+x)^{-1} - e + x\| \leqslant M \|x\|^2$

   for some constant M.

4. X is an algebra and $S = \{x \in X \,|\, xy = yx, \text{ for all } y \in X\}$. This set S is called the *centre* of X. Prove that S is a subalgebra of X, i.e. S is itself an algebra under the algebraic operations in X.

5. X is an algebra and I is a linear subspace of X such that $x \in X$, $i \in I$ imply $xi \in I$. Such a set I is called a *left ideal* in X. Define $x \sim y$ to mean that $x - y \in I$. Prove that $\sim$ is an equivalence relation and that $x_1 \sim x_2$, $y_1 \sim y_2$ imply $x_1 y_1 \sim x_2 y_2$. If E_x is the equivalence class containing x we define $E_x + E_y = E_{x+y}$, $\lambda E_x = E_{\lambda x}$, $E_x E_y = E_{xy}$. Prove that $\{E_x \,|\, x \in X\}$ is an algebra. This algebra is called the *quotient algebra* of X with respect to the ideal I and is denoted by X/I.

6. Let X, Y be algebras and $f : X \to Y$ a homomorphism. Prove that the kernel of f,
   $$\mathrm{Ker}\,(f) = \{x \in X \,|\, f(x) = \theta\},$$
   is a left ideal in X. For the definition of ideal see question 5. Show also that $\mathrm{Ker}\,(f) = \{\theta\}$ if and only if f is an isomorphism.

7. If, in question 6, the homomorphism f is surjective, prove that the quotient algebra $X/\mathrm{Ker}\,(f)$ is isomorphic to Y.

8. (i) X is an algebra and S is the set of all scalar homomorphisms on X. The radical of X is defined as
   $$\mathrm{rad}\,(X) = \cap \{\mathrm{Ker}\,(f) \,|\, f \in S\}.$$

   Prove that $\mathrm{rad}\,(X)$ is an ideal, i.e. $x \in X$, $i \in \mathrm{rad}\,(X)$ imply xi and $ix \in \mathrm{rad}\,(X)$.

   (ii) Let X be the set of 2×2 matrices A such that $a_{21} = 0$. Show that X is an algebra under matrix multiplication and that
   $$\begin{pmatrix} 1 & 0 \\ 0 & 1 \end{pmatrix}, \quad \begin{pmatrix} 0 & 1 \\ 0 & 0 \end{pmatrix}, \quad \begin{pmatrix} 0 & 0 \\ 0 & 1 \end{pmatrix}$$
   is a Hamel base for X. Hence determine the set S of part (i) and show that
   $$\mathrm{rad}\,(X) = \left\{ \begin{pmatrix} 0 & z \\ 0 & 0 \end{pmatrix} \,|\, z \in C \right\}.$$

3. The spectrum and the Gelfand–Mazur theorem

The main aim of this section is to prove a remarkable result of Gelfand and Mazur. This result asserts that the only complex Banach division algebra is the complex algebra C, in the sense that every complex Banach division algebra is isomorphic to C.

In preparation for the proof of this theorem we fix some notation and make certain definitions. Throughout the section, unless we indicate otherwise, X will denote a Banach algebra with identity e. The algebra is always understood to be a complex algebra, though sometimes we shall not emphasize this. Unless specific mention is made to the contrary, X is not necessarily a division algebra.

Units

> $x \in X$ is called a unit of X if and only if x^{-1} exists. By U we denote the set of all units.

Regular point

> Let $x \in X$ be given. Then $\lambda \in C$ is called a regular point of x if and only if $x - \lambda e \in U$.

Spectrum

> The spectrum, $\sigma(x)$, of an element $x \in X$ is the set of all non-regular points of x. Thus $\lambda \in \sigma(x)$ if and only if $x - \lambda e \notin U$.

We should observe that $\sigma(x)$ is a subset of C. In theorem 10 below we shall prove that for each x in a complex Banach algebra with identity we always have $\sigma(x) \neq \varnothing$. For the real case it is possible that there are x such that $\sigma(x) = \varnothing$:

Example 8. Consider the real Banach algebra, with identity, M^2 (see example 5). Let us find $\sigma(A)$, where

$$A = \begin{pmatrix} 0 & -1 \\ 1 & 0 \end{pmatrix}.$$

Now $\sigma(A) = \{\lambda \in R \mid A - \lambda I \text{ has no inverse}\}$. By an elementary theorem of matrix algebra it is known that $A - \lambda I$ has no inverse if and only if $\det(A - \lambda I) = 0$, where det denotes determinant. Hence we have

$\sigma(x) = \{\lambda \in R | \det(A - \lambda I) = 0\}$. But $\det(A - \lambda I) = \lambda^2 + 1$, and there is no real λ such that $\lambda^2 + 1 = 0$. Hence $\sigma(A) = \varnothing$.

It may be known to the reader that the theory of eigenvalues of certain operators is extremely important in Quantum Mechanics. We do not consider it appropriate here to discuss the problems encountered in Quantum Mechanics. However, we shall define the term eigenvalue of an operator on a normed linear space, and show how this notion is related to the spectrum of the operator.

Eigenvalue of an operator

Let X be a normed linear space and A an element of the set $B(X, X)$ of bounded linear operators on X into itself. Then $\lambda \in C$ is called an eigenvalue of A if and only if there is at least one $x \neq \theta$ such that $Ax = \lambda x$. We denote by $\mathrm{Eig}(A)$ the set of all eigenvalues of A.

The set $B(X, X)$ is an algebra with identity, under the usual algebraic operations for operators. Hence we may consider the spectrum $\sigma(A)$ of an operator A in $B(X, X)$.

The general relation between the set of eigenvalues and the spectrum is given by

Theorem 4. $\mathrm{Eig}(A) \subset \sigma(A)$.

Proof. $\lambda \in \mathrm{Eig}(A)$ implies there exists $x_0 \neq \theta$ such that $Ax_0 = \lambda x_0$, i.e. $(A - \lambda E)x_0 = \theta$, where E is the identity operator. Now if $(A - \lambda E)^{-1}$ exists then it is a linear operator, and so

$$(A - \lambda E)^{-1}(A - \lambda E)x_0 = (A - \lambda E)^{-1}\theta = \theta,$$

i.e. $x_0 = \theta$. This contradicts $x_0 \neq \theta$, whence $A - \lambda E$ has no inverse, i.e. $\lambda \in \sigma(A)$. The theorem is thus proved.

The inclusion of theorem 4 may be strict:

Example 9. Consider the operator $A : l_2 \to l_2$, given by

$$Ax = (0, x_1, x_2, \ldots),$$

where $x = (x_1, x_2, \ldots) \in l_2$. Clearly A is linear and bounded. We shall show that $\mathrm{Eig}(A) = \varnothing$, but that $0 \in \sigma(A)$. First, if $0 \in \mathrm{Eig}(A)$ then $(0, x_1, x_2, \ldots) = \theta$, for some $x \neq \theta$, which is impossible. Secondly, if $\lambda \in Eig(A)$, $\lambda \neq 0$, then there exists $x \neq \theta$ such that

$$(0, x_1, x_2, \ldots) = (\lambda x_1, \lambda x_2, \ldots).$$

Hence $0 = x_1$, $x_1 = \lambda x_2, \ldots$, and so $x = \theta$, a contradiction. Thus $\mathrm{Eig}(A) = \varnothing$.

Now $0 \in \sigma(A)$ if and only if A has no inverse. But A has no inverse, for if A^{-1} did exist then $A(A^{-1}x) = x$, for all $x \in l_2$. Putting $x = (1, 0, 0, \ldots)$ we get $A(A^{-1}x) = (1, 0, 0, \ldots)$, whereas the definition of A yields $A(A^{-1}x) = (0, \ldots)$. This contradiction shows that A has no inverse. Hence $\sigma(A)$ contains 0.

There follows a succession of results culminating in the Gelfand–Mazur theorem.

Theorem 5. *If $\|e - x\| < 1$ then $x \in U$.*

Proof. $\sum_0^\infty \|(e - x)^k\| < \infty$ and so, writing $x = e - (e - x)$, we have

$$x\left(e + \sum_0^\infty (e - x)^k\right) = e,$$

i.e. the inverse of x is $e + \Sigma(e - x)^k$.

Corollary. *If $\lambda \in C$ and $\|x\| < |\lambda|$, then $x - \lambda e \in U$.*

Theorem 6. *U is an open subset of X.*

Proof. Let $x \in U$ and write $S(x) = \{y \mid \|x - y\| < \|x^{-1}\|^{-1}\}$. Then $\|e - yx^{-1}\| < 1$, whenever $y \in S(x)$.

It follows by theorem 5 that $yx^{-1} \in U$ when $y \in S(x)$. Now y has inverse $x^{-1}(yx^{-1})^{-1}$ when $y \in S(x)$. Hence $y \in S(x)$ implies $y \in U$, whence U is open.

Theorem 7. *For each $x \in X$, the spectrum $\sigma(x)$ is a compact subset of C.*

Proof. By the Heine–Borel theorem it is enough to show that $\sigma(x)$ is bounded and closed. Take $\lambda \in \sigma(x)$. Then $x - \lambda e \notin U$ and the corollary to theorem 5 yields $\|x\| \geqslant |\lambda|$. Hence $\sigma(x)$ lies in the disc centre $(0, 0)$ and radius $\|x\|$.

Now let $\lambda_0 \in \sim \sigma(x)$. Then $x - \lambda_0 e \in U$. Write $x - \lambda e = f(\lambda)$. By theorem 6 there exists $S(f(\lambda_0)) \subset U$. Let $\lambda \to \lambda_0$. Then $f(\lambda) \to f(\lambda_0)$ and so there exists a neighbourhood $N(\lambda_0)$ such that $\lambda \in N(\lambda_0)$ implies $f(\lambda) \in S(f(\lambda_0))$. Hence $\lambda \in N(\lambda_0)$ implies $f(\lambda) \in U$, whence $\lambda \in \sim \sigma(x)$. Thus $\sim \sigma(x)$ is open, and so $\sigma(x)$ is closed.

Theorem 8. *Let* $y, y_0 \in U$. *Then* $y \to y_0$ *implies*

$$\text{(i)} \ y^{-1}y_0 \to e, \qquad \text{(ii)} \ y^{-1} \to y_0^{-1}.$$

Proof. (i) Let $\epsilon > 0$ be given. If $y \to y_0$ then $y_0^{-1}y \to e$, and so there exists δ_1 such that $\|y - y_0\| < \delta_1$ implies $\|y_0^{-1}y - e\| < 2^{-1}$. By theorem 5 we have

$$(y_0^{-1}y)^{-1} = e + \Sigma(e - y_0^{-1}y)^k,$$

$$y^{-1}y_0 - e = \Sigma(e - y_0^{-1}y)^k.$$

Hence, when $\|y - y_0\| < \delta_1$, we have

$$\|y^{-1}y_0 - e\| < \sum_{1}^{n} \|e - y_0^{-1}y\|^k + \sum_{n+1}^{\infty} 2^{-k}.$$

Choose n large so that $\sum_{n+1}^{\infty} 2^{-k} < \epsilon$ and then choose $\delta_2 = \delta_2(\epsilon)$ so small that $\|y - y_0\| < \delta_2$ implies that the finite sum

$$\sum_{1}^{n} \|e - y_0^{-1}y\|^k$$

is $< \epsilon$. This is possible since $y \to y_0$ implies $y_0^{-1}y \to e$. Hence we have $\|y^{-1}y_0 - e\| < 2\epsilon$ if $\|y - y_0\| < \min(\delta_1, \delta_2)$, which proves (i).

   (ii) $y \to y_0$ implies $y^{-1}y_0 \to e$, by (i). Hence

$$\|(y^{-1}y_0) y_0^{-1} - y_0^{-1}\| \leqslant \|y_0^{-1}\| \, \|y^{-1}y_0 - e\| \to 0,$$

i.e. $y^{-1} \to y_0^{-1}$ as $y \to y_0$.

Theorem 9. *For each* $x \in X$, $x(\lambda) = (x - \lambda e)^{-1}$ *is analytic on* $\sim \sigma(x)$.

Proof. If $\lambda \in \sim \sigma(x)$ then $x - \lambda e \in U$, so $x(\lambda)$ is well-defined. By theorem 7, $\sim \sigma(x)$ is open. Now let $\lambda, \lambda_0 \in \sim \sigma(x)$. Then it is easy to check that

$$x(\lambda_0) = x(\lambda) + (\lambda_0 - \lambda) x(\lambda) x(\lambda_0).$$

By theorem 8 (ii), $\lambda \to \lambda_0$ implies $x(\lambda) \to x(\lambda_0)$, whence

$$\frac{x(\lambda) - x(\lambda_0)}{\lambda - \lambda_0} \to \{x(\lambda_0)\}^2 \quad \text{as} \quad \lambda \to \lambda_0.$$

Thus $x(\lambda)$ is analytic in the sense defined in section 5, chapter 4.

   At this stage we remind the reader that X is a complex Banach algebra with identity.

Theorem 10. *For each* $x \in X$, $\sigma(x) \neq \varnothing$.

Proof. If $\sigma(x) = \varnothing$ then $x - \lambda e$ has an inverse for all λ, i.e.

$$x(\lambda) = (x - \lambda e)^{-1}$$

exists for all λ. By theorem 9, $x(\lambda)$ is an integral function. Now if $\lambda \neq 0$ then

$$\|(x - \lambda e)^{-1}\| = |\lambda|^{-1} \|(e - x/\lambda)^{-1}\|$$

and so $x(\lambda) \to \theta$ as $|\lambda| \to \infty$. Thus $x(\lambda)$ is bounded on C. By the generalized Liouville theorem (theorem 20, section 5, chapter 4) we have that $x(\lambda) = x(0)$, for all λ. Letting $|\lambda| \to \infty$ we get $x(0) = \theta$, whence $x(\lambda) = \theta$, for all λ. It follows that $(x - \lambda e) x(\lambda) = \theta$, i.e. $e = \theta$, contrary to the fact that X is an algebra with identity $e \neq \theta$. Hence $\sigma(x) \neq \varnothing$.

We may now prove the Gelfand–Mazur theorem.

Theorem 11. *Let X be a complex Banach division algebra. Then X is isomorphic to C.*

Proof. $x \in X$ implies $\sigma(x) \neq \varnothing$, by theorem 10. Hence there exists $\lambda \in \sigma(x)$, so $x - \lambda e$ has no inverse. Since X is a division algebra it follows that $x - \lambda e = \theta$, i.e. $x = \lambda e$. Thus, to each $x \in X$ corresponds a $\lambda \in C$ such that $x = \lambda e$. This λ is unique. For $x = \lambda e$, $x = \mu e$, $\alpha = \lambda - \mu \neq 0$ imply $\alpha e = \theta$, and so $e = \theta$, a contradiction.

To summarize, we have a map $f : X \to C$, given by $f(x) = \lambda$, where $x = \lambda e$. Now if $x = \lambda e$, $y = \mu e$, then $x + y = (\lambda + \mu)e$, $xy = (\lambda \mu)e$, and so f is linear and multiplicative. If $f(x) = f(y)$ then $\lambda = \mu$ and so $x = y$. Finally, if $\lambda \in C$ is given, then the element $x = \lambda e \in X$ is such that $f(x) = \lambda$. Hence f is bijective.

This proves the theorem.

Corollary. *Let X be a complex Banach division algebra. Then*

    (i) *X is necessarily commutative.*

    (ii) *X is necessarily one-dimensional.*

Proof. (i) Let $x, y \in X$. Then $x = \lambda e$, $y = \mu e$ and so

$$xy = (\lambda \mu)e = (\mu \lambda)e = yx.$$

    (ii) The set $\{e\}$ is, clearly, a Hamel base for X.

Exercises 3

1. Let X be an algebra with identity. Show that the set U of units is a multiplicative group.
2. Let $A \in M^n$ (see example 5). Prove that $\sigma(A)$ is never empty but has at most n elements.

3. Let

$$A = \begin{pmatrix} a & b \\ c & d \end{pmatrix} \in M^2.$$

Prove that $\sigma(A)$ is a one-point set if and only if $(a+d)^2 = 4 \det(A)$.

4. For any set S in an algebra, let $S^n = \{a^n | a \in S\}$. If X is a Banach algebra with identity, prove that

$$\sigma(x)^n = \sigma(x^n) \quad (n = 2, 3, \ldots).$$

As a hint, observe that $x^n - z^n e = (x - ze)(x^{n-1} + \ldots + z^{n-1} e)$ for each complex z.

5. With X as in question 4, write $r(x) = \sup\{|\lambda| \,|\, \lambda \in \sigma(x)\}$. Then $r(x)$ is called the spectral radius of x. Prove that $0 \leqslant r(x) \leqslant \|x\|$. Using $\sigma(x)^n = \sigma(x^n)$ show that $r(x^n) = \{r(x)\}^n$ and deduce that

$$r(x) \leqslant \lim \inf_n \|x^n\|^{1/n}.$$

6. Suppose that $r(x)$ in question 5 is positive. Prove that

$$\sum_1^\infty \lambda^{-k} x^{k-1}$$

diverges for $|\lambda| < r(x)$ and converges for $|\lambda| > r(x)$. Deduce that

$$r(x) = \lim \sup_n \|x^n\|^{1/n}.$$

Show also that this formula holds for $r(x) = 0$.

Combine questions 5 and 6 to show that

$$r(x) = \lim_n \|x^n\|^{1/n}.$$

7. Let $A : l_1 \to l_1$ be defined by $Ax = (x_2, x_3, \ldots)$, where $x = (x_1, x_2, \ldots) \in l_1$. Prove that A is a bounded linear operator, with $\|A\| = 1$.

In the context of the algebra $B(l_1, l_1)$ show that $|\lambda| > 1$ implies $\lambda \notin \sigma(A)$, and that $|\lambda| \leqslant 1$ implies $\lambda \in \mathrm{Eig}(A)$. Deduce that the spectrum of A is the closed unit disc in the complex plane.

8. Find the spectrum of the operator $A : l_2 \to l_2$, given by $Ax = (0, x_1, x_2, \ldots)$, for $x = (x_1, x_2, \ldots) \in l_2$. The underlying algebra is $B(l_2, l_2)$.

9. Let $A : l_1 \to l_1$ be defined by the infinite matrix

$$A = \begin{bmatrix} 0 & 1 & 1 & 1 & . & . & . \\ 1 & 0 & 0 & 0 & . & . & . \\ 0 & 1 & 0 & 0 & . & . & . \\ 0 & 0 & 1 & 0 & . & . & . \\ . & . & . & . & . & . \end{bmatrix}.$$

Thus $Ax = (x_2 + x_3 + x_4 + \ldots, x_1, x_2, x_3, \ldots)$, where $x = (x_1, x_2, \ldots)$. Let $D = \{\lambda | \,|\lambda| \leqslant 1\}$ and $\lambda_0 = (1 + \sqrt{5})/2$. Prove that $\sigma(A) = D \cup \{\lambda_0\}$. The underlying algebra is $B(l_1, l_1)$.

4. The Gelfand representation theorem

Throughout this section we shall suppose that X is a commutative Banach algebra with identity e. Our object is to prove a weak version of a fundamental theorem of Gelfand, which tells us that a certain type of Banach algebra (called semisimple) is isomorphic to some function algebra.

First we define some simple concepts which are used in the proof of Gelfand's theorem. A set $I \subset X$ is called an ideal if and only if I is a linear subspace of X and $x \in X$, $y \in I$ imply $xy \in I$. If the inclusion $I \subset X$ is strict then I is called a proper ideal.

An ideal M is called maximal if and only if (i) M is a proper ideal and (ii) whenever $I \supset M$ properly, where I is an ideal, then $I = X$. Since we assume X has an identity, we have $X \neq \{\theta\}$. Thus $\{\theta\}$ is a proper ideal. By using Zorn's lemma, as in previous chapters, it is easy to show that there exists a maximal ideal M such that $\{\theta\} \subset M$. Hence every X contains a maximal ideal, so that it makes sense to speak of $\hat{M}$, the set of all maximal ideals.

X is called semisimple if and only if

$$\bigcap_{M \in \hat{M}} M = \{\theta\}.$$

We denote by $F(\hat{M})$ the set of all bounded complex valued functions on $\hat{M}$. With pointwise addition and multiplication $F(\hat{M})$ becomes a commutative Banach algebra with identity. We now state our version of the Gelfand representation theorem.

Theorem 12. *Let X be a semisimple commutative Banach algebra with identity. Then X is isomorphic to a subalgebra of the function algebra $F(\hat{M})$.*

Proof. Let $M \in \hat{M}$ and consider the quotient algebra X/M. As in question 5, exercises 2, of this chapter, we may define X/M, merely replacing the I there by M. Since M is a maximal ideal it turns out that X/M is a division algebra. To show this we take any nonzero element E in X/M, i.e. take $E \neq M$. Take $x \in E$ and consider

$$I = \{xy - m \mid y \in X, m \in M\}.$$

Then $I \supset M$ strictly and I is an ideal. Since M is maximal we must have $I = X$, whence $e \in I$, so that $xy - m = e$ for some $y \in X$

and some $m \in M$. Thus, if E_y is the equivalence class containing y, then

$$E_x . E_y = E_{m+e} = \{z | z - (m + e) \in M\}$$
$$= \{z | z - e \in M\} = E_e.$$

Hence E_y is the inverse of $E = E_x$, so that X/M is a division algebra. By defining

$$\|E_x\| = \inf\{\|y\| \, | y \in E_x\}$$

we turn X/M into a normed space (see chapter 4, section 1). Now the closure $\bar{M}$ of M is a proper ideal. For it is trivial that $\bar{M}$ is an ideal, and if $x \in M$ then x^{-1} does not exist. If it did then $e \in M$, whence $M = X$, a contradiction. Thus $M \subset {\sim} U$, whence $\bar{M} \subset \overline{{\sim} U} = {\sim} U$, since U is open by theorem 6. But $e \notin {\sim} U$, so that $e \notin \bar{M}$, whence $\bar{M}$ is a proper ideal. We now have that $M = \bar{M}$, for if $M \subset \bar{M}$ strictly then $\bar{M} = X$, contrary to the fact that $\bar{M}$ is proper. Thus M is a closed maximal ideal. By theorem 3, chapter 4, it follows that X/M is a Banach division algebra. The Gelfand–Mazur theorem (theorem 11), now shows that X/M is isomorphic to C.

Our next task is to exhibit an isomorphism between X and a subalgebra of $F(\hat{M})$. Take $x \in X$ and $M \in \hat{M}$. Let $E \in X/M$ be the equivalence class containing x. By the Gelfand–Mazur theorem, $E = \lambda E_e$, where $\lambda = \lambda(x, M) \in C$. Denote by $\hat{x}$ that function on $\hat{M}$ defined by $\hat{x}(M) = \lambda(x, M)$. Thus we have a mapping $f : X \to F(\hat{M})$ given by $f(x) = \hat{x}$. By the Gelfand–Mazur theorem we know that λ, as a function of x, is a homomorphism. Thus, for example,

$f(x) + f(y) = \hat{x} + \hat{y} = f(x+y)$, since $\hat{x}(M) + \hat{y}(M) = \widehat{x + y}(M)$ for each M. Now f is injective. For if $f(x) = f(y)$ then $\lambda(x - y, M) = 0$ for all M, so that $E_{x-y} = 0$. $E_e = M$ for all M. Thus $x - y \in \cap \{M | M \in \hat{M}\}$ and since X is semisimple we have $x - y = \theta$, whence f is injective.

It is now clear that the image of X under f is a subalgebra of the algebra of all complex valued functions on $\hat{M}$. Finally, $\hat{x}$ is bounded on $\hat{M}$, for $E_x(M) = \lambda(x, M) E_e(M)$, which implies

$$|\hat{x}(M)| \, \|E_e(M)\| \leqslant \|E_x(M)\|.$$

But
$$\|E_x(M)\| \leqslant \|x\| \quad \text{and} \quad \|E_e(M)\| \geqslant 1,$$

whence $|\hat{x}(M)| \leqslant \|x\|$ for each M. Thus $\hat{x}$ is bounded on M, and this completes the proof of the theorem.

We remark that we have proved only a weak version of the Gelfand representation theorem. The strong form of the theorem asserts that every semisimple commutative Banach algebra with identity is isomorphic to a subalgebra of the algebra of continuous complex valued functions on the compact Hausdorff space $\hat{M}$. To prove this result we need more topological results than we have at our disposal, and the interested reader is referred to more advanced works on Banach algebras for the details.

Exercises 4

1.   (i) Let X be a commutative algebra with identity and suppose that I is a proper ideal in X. Partially order, by set inclusion, the class of all proper ideals in X, and then use Zorn's lemma to prove that there is a maximal ideal M which contains I.

     (ii) Let $x \in\, \sim U$. Show that $I = \{xy | y \in X\}$ is a proper ideal in X. Use part (i) to show that there is a maximal ideal which contains x.

2. Suppose X is a commutative Banach algebra with identity e. In the notation of the Gelfand theorem, prove that $\lambda(e, M) = 1$, for each $M \in \hat{M}$, and $\lambda(x, M) = 0$ if and only if $x \in M$.

3.   (i) Let X be a commutative Banach algebra with identity. Write $I = \cap\, \{M | M \in \hat{M}\}$. Prove that I is a closed ideal and that X/I is semisimple.

     (ii) Prove that $x \in U$ if and only if $\lambda(x, M) \neq 0$ for any $M \in \hat{M}$. As a hint use 1 (ii) and the second part of 2.

6

HILBERT SPACE

1. Inner product and Hilbert spaces

In this chapter we make a brief excursion into Hilbert space. Modern developments in Hilbert space are concerned largely with the theory of operators on the space, rather than with the space itself. We shall not attempt to discuss any operator theory in this introductory work. All we wish to do is to define such basic ideas as inner product space, Hilbert space, orthonormal set and prove some classical theorems (Riesz–Fischer, Parseval, Riesz–Fréchet, and the result that l_2 is essentially the only infinite dimensional separable Hilbert space).

The theory of Hilbert space may be said to have started in 1912 with Hilbert's 'Grundzuge einer allgemeinen Theorie der linearen Integralgleichungen'.‡ However, several years elapsed before an axiomatic basis was provided by the famous J. von Neumann (1903–1957).

The fundamental underlying structure is the

Inner product space

> An inner product space X (or pre-Hilbert space) is a complex linear space together with an inner product $(\ ,\ ): X \times X \to C$, such that (i) $(x, y) = \overline{(y, x)}$; (ii) $(\lambda x + \mu y, z) = \lambda(x, z) + \mu(y, z)$, (iii) $(x, x) \geqslant 0$, with equality only for $x = \theta$.

In (i) the bar denotes the complex conjugate of (y, x). In (ii), x, y, z are in X and λ, μ are in C. There are other notations for inner products, such as $\langle x, y \rangle$, $(x|y)$, but we shall not use them. Also, some use z^* instead of $\bar{z}$ for complex conjugates.

From (i) and (ii) we deduce that $(x, \lambda y + \mu z) = \bar{\lambda}(x, y) + \bar{\mu}(x, z)$. For any x, y in X and any λ in C we thus have

$$0 \leqslant (x - \lambda y, x - \lambda y) = (x, x) - 2Rl[\lambda(y, x)] + |\lambda|^2 (y, y).$$

Suitable choice of λ then yields

$$|(x, y)| \leqslant (x, x)^{\frac{1}{2}} (y, y)^{\frac{1}{2}} \quad \text{[Schwarz' inequality]}.$$

† D. Hilbert (1862–1943), the great German mathematician, made important contributions to mathematical logic, analysis, geometry and mathematical physics.

Theorem 1. *Each inner product space is a normed linear space under* $\|x\| = +\sqrt{(x, x)}$.

Proof. $\|x\| = 0$ implies $x = \theta$, by (iii) of the definition of inner product space.
$$(\lambda x, \lambda x) = \lambda \bar{\lambda}(x, x) = |\lambda|^2 (x, x),$$
whence $\|\lambda x\| = |\lambda| \, \|x\|$.
$$(x + y, x + y) = (x, x) + 2Rl[(x, y)] + (y, y),$$
whence, by Schwarz' inequality, $\|x + y\|^2 \leqslant \|x\|^2 + 2\|x\| \cdot \|y\| + \|y\|^2$, so that $\|x + y\| \leqslant \|x\| + \|y\|$. We have thus shown that an inner product generates a norm, $\|x\| = +\sqrt{(x, x)}$.

Hilbert space

A Hilbert space H is a complete inner product space, i.e. a Banach space whose norm is generated by an inner product.

Isomorphic inner product spaces

Inner product spaces X, X' are called isomorphic if and only if there exists $f: X \to X'$ such that f is linear, bijective and satisfies $(f(x), f(y)) = (x, y)$ for all $x, y \in X$.

In the definition of isomorphic spaces we have used the same notation for the inner product in X' as in X. Since a Hilbert space is an inner product space, the notion of isomorphism applies there.

We remark that our definition of Hilbert space allows finite dimensional spaces. Some authors require that a Hilbert space should be infinite dimensional, since finite dimensional spaces are not so interesting. When dimension looms we say so explicitly.

Example 1. (i) C^n is an n-dimensional Hilbert space, under the inner product
$$(x, y) = \sum_{k=1}^{n} x_k \bar{y}_k.$$

(ii) $C[0, 1]$ is an infinite dimensional inner product space, under
$$(x, y) = \int_0^1 x(t) \, \overline{y(t)} \, dt.$$

It is not a Hilbert space, since it is not complete under $\|x\| = \sqrt{(x, x)}$.

(iii) $L_2[0, 1]$, with

$$(x, y) = \int_0^1 x(t)\,\overline{y(t)}\,dt,$$

is an infinite dimensional Hilbert space. Completeness was proved in section 1, chapter 4.

(iv) Let A be an index set of elements. Let $x : A \to C$. Then we say that $x \in l_2(A)$ if and only if

$$\sup \Sigma |x(\alpha)|^2 < \infty,$$

where the supremum is taken over all finite sums of the form $|x(\alpha_1)|^2 + \ldots + |x(\alpha_n)|^2$, where $\alpha_1, \ldots, \alpha_n$ are distinct elements of A.

$l_2(A)$ is a Hilbert space under

$$(x, y) = \sup \Sigma x(\alpha)\,\overline{y(\alpha)}.$$

(v) $l_2 = \{x \mid \Sigma |x_k|^2 < \infty\}$ is the space originally studied by Hilbert. It is a Hilbert space under

$$(x, y) = \sum_{k=1}^{\infty} x_k \overline{y}_k.$$

It is clear that l_2 can be regarded as $l_2(N)$ of (iv), where N denotes the positive integers.

Orthonormal sets

Let X be an inner product space. We say that x is orthogonal to y (written $x \perp y$) if and only if $(x, y) = 0$. A set $S \subset X$ is called orthogonal if and only if $x \perp y$ for any distinct vectors $x, y \in S$. If, in addition, $\|x\| = 1$, for all $x \in S$, then S is called an orthonormal set.

Example 2. (i) Write $e_k = (0, 0, \ldots, 1, \ 0, \ \ldots)$, with 1 in the kth place. Then $S = \{e_1, e_2, \ldots\}$ is orthonormal in l_2.

(ii) $S = \{e^{ikt}/\sqrt{2\pi} \mid k = 0, \pm 1, \pm 2, \ldots\}$ is orthonormal in $L_2[0, 2\pi]$.

(iii) Any orthogonal set S of non-zero vectors in an inner product space is linearly independent. For $\Sigma \lambda_k s_k = \theta$ implies

$$0 = (\Sigma \lambda_k s_k, \Sigma \lambda_k s_k) = \Sigma |\lambda_k|^2 \|s_k\|^2,$$

whence $\lambda_k = 0$ for the relevant k.

(iv) Any orthogonal set S of non-zero vectors may be normalized: $T = \{s\|s\|^{-1} \mid s \in S\}$ is obviously orthonormal.

Any orthonormal sequence is linearly independent, by example 2 (iii). Given a linearly independent sequence, we may construct from it an orthonormal sequence, by use of

Theorem 2 (Gram–Schmidt orthonormalization). *Let $S = \{s_1, s_2, \ldots\}$ be a linearly independent sequence in an inner product space. Then there is an orthonormal sequence $T = \{t_1, t_2, \ldots\}$ such that*

$$\text{l. hull}\,(S) = \text{l. hull}\,(T).$$

Proof. Since S is linearly independent, all s_k are non-zero. Define $t_1 = s_1 \|s_1\|^{-1}$, so that $\|t_1\| = 1$. Define $v = s_2 - (s_2, t_1) t_1$. Then $v \perp t_1$ and $v \neq \theta$, since $\{s_1, s_2\}$ is linearly independent. Hence $t_2 = v \|v\|^{-1}$ is orthogonal to t_1 and $\|t_2\| = 1$. In general we proceed inductively, defining

$$v = s_n - \sum_1^{n-1} (s_n, t_k) t_k \quad \text{and} \quad t_n = v \|v\|^{-1}.$$

It is clear from the construction that S and T have the same linear hull.

Theorem 3. *A finite dimensional inner product space X is necessarily a Hilbert space.*

Proof. Let $S = \{s_1, \ldots, s_n\}$ be a Hamel base. By theorem 2 there exists an orthonormal (and so linearly independent) set $T = \{t_1, \ldots, t_n\}$, such that l. hull $(T) = $ l. hull (S). Since l. hull $(S) = X$, T is an orthonormal Hamel base for X. Hence $x \in X$ implies $x = \Sigma \lambda_k t_k$, $\|x\|^2 = \Sigma |\lambda_k|^2$, the sums being over $1 \leqslant k \leqslant n$. A trivial argument now shows that X is complete.

Exercises 1

1. Prove that $\|x+y\|^2 + \|x-y\|^2 = 2(\|x\|^2 + \|y\|^2)$, in any inner product space. This result is known as the *parallelogram law*.

   Show also that

   $$4(x, y) = \|x+y\|^2 - \|x-y\|^2 + i\,\|x+iy\|^2 - i\,\|x-iy\|^2.$$

2. Let X be a Banach space in which the parallelogram law holds (see 1, above). Use question 1 to show that X may be made into a Hilbert space.
3. In an inner product space, suppose $y \neq \theta$. Prove that

   $$\|x+y\| = \|x\| + \|y\| \quad \text{if and only if} \quad y = px$$

   for some real $p \geqslant 0$.
4. Find the necessary and sufficient condition for equality to hold in the Schwarz' inequality.

5. Prove that the set S of example 2 (ii) is orthogonal in $L_2[0, 2\pi]$.
6. In the Gram–Schmidt process (theorem 2), let S consist of the functions $1, t, t^2, \ldots$ in the space $L_2[-1, 1]$. Show that the orthonormal set T consists of the normalized Legendre polynomials

$$\frac{1}{\sqrt{2}}, \quad \sqrt{\frac{3}{2}} t, \quad \frac{1}{2} \sqrt{\frac{5}{2}} (3t^2 - 1), \quad \ldots.$$

7. Let X be linear. Then X has a Hamel base B, by theorem 4, section 2, chapter 3. If $x = \Sigma \lambda_k b_k$, $y = \Sigma \mu_k b_k$, show that $(x, y) = \Sigma \lambda_k \bar{\mu}_k$ is an inner product for X.

8. Prove that $(x, y) = \sum_1^\infty x_k \bar{y}_k$ is an inner product for l_p, when $0 < p \leqslant 2$. Is it an inner product when $p > 2$?

9. Let X be an inner product space. Prove that $x \perp y$ is equivalent to $y \perp x$, and $x \perp x$ is equivalent to $x = \theta$.

   Prove that $x \perp \{y_1, \ldots, y_n\}$ implies $x \perp \Sigma \lambda_k y_k$. Does $x \perp y$ and $y \perp z$ imply $x \perp z$?

10. Let X be an inner product space.
    (i) If $x_n \to x$, $y_n \to y$ (in norm), show that $(x_n, y_n) \to (x, y)$ (in modulus).
    (ii) If (x_n), (y_n) are Cauchy sequences (in norm), show that $((x_n, y_n))$ is convergent.

11. Let $p \geqslant 1$. An inner product is introduced into l_p in such a way that it generates the usual norm. Prove that $p = 2$.

2. Orthonormal sets

We defined the term orthonormal set in section 1. In this section we shall assume throughout that our orthogonal, or orthonormal, sets are in a Hilbert space H. It will be obvious that some of our results still hold in any inner product space. We shall prove the classical results of Riesz–Fischer and Parseval, which lead to the identification of all infinite dimensional separable Hilbert spaces.

Theorem 4. *If $\{x_1, \ldots, x_n\}$ is orthogonal, then*

$$\|\Sigma x_k\|^2 = \Sigma \|x_k\|^2.$$

Proof. $(x_1 + \ldots + x_n, x_1 + \ldots + x_n) = (x_1, x_1) + \ldots + (x_n, x_n).$

Theorem 5. *Let (x_k) be an orthogonal sequence. Then Σx_k converges if and only if $\Sigma \|x_k\|^2 < \infty$.*

Proof. Write $s_n = x_1 + \ldots + x_n$, so that Σx_k converges if and only if (s_n) is Cauchy. Now, by theorem 4,

$$\|s_{n+p} - s_n\|^2 = \left\| \sum_{n+1}^{n+p} x_k \right\|^2 = \sum_{n+1}^{n+p} \|x_k\|^2,$$

whence the result.

Fourier coefficient

Let (e_k) be an orthonormal sequence and let $x \in H$. Then $\alpha_k = (x, e_k)$ is called the Fourier coefficient of x relative to e_k.

Example 3. Take (e_k) as in example 2 (i), section 1. Then

$$\alpha_k = (x, e_k) = x_k, \quad \text{for any} \quad x \in l_2.$$

We observe here that the sequence (α_k) of Fourier coefficients is in l_2. This is true in any Hilbert space:

Theorem 6 (Bessel's inequality). *Let (e_k) be an orthonormal sequence and let $x \in H$. Then $(\alpha_k) \in l_2$ and $\|(\alpha_k)\| \leqslant \|x\|$, where (α_k) is the sequence of Fourier coefficients of x and $\|(\alpha_k)\|$ denotes the l_2 norm.*

Proof. Let us approximate to x by $\sum\limits_{k=1}^{n} \alpha_k e_k$. Precisely, consider

$$y = x - \sum_{k=1}^{n} \alpha_k e_k.$$

Then for $1 \leqslant p \leqslant n$,

$$(y, e_p) = (x, e_p) - \sum_{k=1}^{n} \alpha_k(e_k, e_p) = 0,$$

since (e_k) is orthonormal. Thus $y \perp e_p$ and by theorem 4,

$$\|x\|^2 = \|y\|^2 + \sum_{k=1}^{n} \|\alpha_k e_k\|^2 \geqslant \sum_{k=1}^{n} |\alpha_k|^2.$$

Bessel's inequality follows on letting $n \to \infty$.

The following classical theorem is a kind of converse to the previous result.

Theorem 7 (Riesz–Fischer). *Let (e_k) be an orthonormal sequence and let (β_k) be an arbitrary sequence in l_2. Then there exists $x \in H$ such that*

$$\beta_k = (x, e_k) \quad \text{and} \quad \|(\beta_k)\| = \|x\|,$$

where $\|(\beta_k)\|$ denotes the l_2 norm.

Proof. $\Sigma \|\beta_k e_k\|^2 = \Sigma |\beta_k|^2 < \infty$, so $\Sigma \beta_k e_k$ converges to some $x \in H$, by theorem 5. Now if $n > p$,

$$\left(x - \sum_1^n \beta_k e_k, e_p\right) = (x, e_p) - \beta_p.$$

By Schwarz' inequality we get

$$|(x, e_p) - \beta_p| \leqslant \left\|x - \sum_1^n \beta_k e_k\right\| \to 0 \quad (n \to \infty),$$

so that $\beta_p = (x, e_p)$ for each p. Finally,

$$\sum_1^n |\beta_k|^2 = \left\|\sum_1^n \beta_k e_k\right\|^2.$$

Letting $n \to \infty$, we obtain $\|(\beta_k)\|^2 = \|x\|^2$, which proves the theorem.

The theorem that has just been proved, in the context of a general Hilbert space, was originally proved in 1907 by Riesz and Fischer for the special Hilbert space $L_2[0, 2\pi]$ with its standard orthonormal set $\{e^{ikt}/\sqrt{2\pi}\}$. It is to be noted that completeness of the space is implicit in theorem 7, where it is merely part of the hypothesis. In the original Riesz–Fischer theorem the completeness of $L_2[0, 2\pi]$, which is quite non-trivial (see section 1, chapter 4), had to be proved. Thus the classical result is much more of an achievement than our theorem 7 would superficially indicate.

We define now the important concept of

Total set

A set S is called total in H if and only if, $x \in H$ and $(x, s) = 0$ for all $s \in S$, imply $x = \theta$.

Example 3. $S = \{e_k\}$ is total in l_2. For $(x, e_k) = 0$ for all k implies $x_k = 0$ for all k, i.e. $x = \theta$.

Theorem 8 (Parseval's theorem). *Let (e_k) be an orthonormal sequence. Then (e_k) is total, if and only if, for each $x \in H$, $\|x\|^2 = \Sigma |\alpha_k|^2$, where $\alpha_k = (x, e_k)$.*

Proof. If $\|x\|^2 = \Sigma |\alpha_k|^2$ and $(x, e_k) = 0$ for all k, then $\|x\|^2 = 0$, whence $x = \theta$. Conversely, suppose (e_k) is total. Let $x \in H$. Then $(\alpha_k) \in l_2$ by theorem 6. By the Riesz–Fischer theorem (theorem 7), there exists $y \in H$ such that $\alpha_k = (y, e_k)$ and $\|y\|^2 = \Sigma |\alpha_k|^2$. Hence

$$(x - y, e_k) = \alpha_k - (y, e_k) = 0,$$

for all k. Since (e_k) is total we have $x = y$. The result follows.

Corollary. *If (e_k) is a total orthonormal sequence then every $x \in H$ has the Fourier expansion $x = \Sigma \alpha_k e_k$, where $\alpha_k = (x, e_k)$.*

Proof. By theorem 7 and theorem 8, if $x \in H$ then $x = \Sigma \alpha_k e_k$.

Theorem 9. *Let H be a separable infinite dimensional Hilbert space. Then H is isomorphic to l_2.*

Proof. There exists $S = \{s_k\}$ dense in H. Let $x \in H$. Then $s_{k_i} \to x$ for some sequence (s_{k_i}). Hence $(x, s_k) = 0$ for all $s_k \in S$ implies

$$\lim_i (x, s_{k_i}) = (x, x) = 0,$$

so that $x = \theta$. Thus S is total.

From any given set in a linear space we may determine, using Zorn's lemma, a linearly independent subset whose linear hull is the linear hull of the given set (employ the argument of theorem 4, section 2, chapter 3). Hence we may determine a linearly independent subset $T = \{t_k\}$ of S such that $\mathrm{l.\,hull}\,(T) = \mathrm{l.\,hull}\,(S)$. Now if $(x, t_k) = 0$ for all $t_k \in T$, then $(x, s_k) = (x, \Sigma \lambda_i t_i) = \Sigma \overline{\lambda}_i (x, t_i) = 0$, for every k. Hence $x = \theta$, and so T is total. By theorem 2 we orthonormalize T to produce a set $E = \{e_k\}$, which is seen to be total by the same argument that showed that T was total.

By the corollary to theorem 8, every $x \in H$ may be written as $x = \Sigma \alpha_k e_k$, from which it follows that E is not finite.

Now define $f : H \to l_2$ by $f(x) = (\alpha_k)$, where $\alpha_k = (x, e_k)$. Then f is well-defined, injective and surjective, by Bessel's inequality, totality of E and the Riesz–Fischer theorem, respectively. Obviously, f is linear. Finally, by the corollary to theorem 8, if $x, y \in H$ then $x = \Sigma \alpha_k e_k$, $y = \Sigma \beta_k e_k$, where α_k, β_k are, respectively, the Fourier coefficients of x, y. Now

$$\left(\sum_1^n \alpha_k e_k, \sum_1^n \beta_k e_k \right) = \sum_1^n \alpha_k \overline{\beta}_k,$$

whence, letting $n \to \infty$, we get $(x, y) = \Sigma \alpha_k \bar{\beta}_k$. But $\Sigma \alpha_k \bar{\beta}_k = ((\alpha_k), (\beta_k))$, with the inner product of l_2. Hence $(x, y) = (f(x), f(y))$, so that we have now shown that f is a Hilbert space isomorphism. This proves the theorem.

Exercises 2

1. Let (x_k) be an orthogonal sequence in a Hilbert space and suppose that $(\|x_k\|) \in l_p$, where $0 < p < 2$. Prove that Σx_k converges. Can orthogonality be relaxed?

2. Let (e_k) be an orthogonal sequence in an inner product space.
   Prove that
   (i) the sequence (α_k) of Fourier coefficients is in c_0,
   (ii) for any $\lambda_1, \ldots, \lambda_n$,

   $$\left\| x - \sum_1^n \lambda_k e_k \right\|^2 = \|x\|^2 - \sum_1^n |\alpha_k|^2 + \sum_1^n |\alpha_k - \lambda_k|^2.$$

   Deduce that the Fourier coefficients give the best approximation to x by a linear combination of $e_1, \ldots, e_n$.

3. Suppose (e_k) is orthonormal in H and that (λ_k) is a sequence of scalars such that $\Sigma \lambda_k e_k$ converges, to x say. Prove that λ_k must be the Fourier coefficient of x relative to e_k.

4. Let (e_k) be orthonormal in H. Prove that (e_k) is total if and only if $x = \Sigma \alpha_k e_k$, for each $x \in H$.

5. Prove that every n-dimensional inner product space is isomorphic to C^n.

6. (i) It is shown in books on Lebesgue integration that $L_2[0, 2\pi]$ is separable. Use this result to prove that $L_2[0, 2\pi]$ is isomorphic to l_2.
   (ii) It may be proved, using the classical theorem of Fejér on the Cesàro summability of Fourier series, that if $x \in L_2[0, 2\pi]$ then

   $$\int_0^{2\pi} |x(t)|^2 \leqslant \sum_{-\infty}^{\infty} |\alpha_k|^2,$$

   where $$\alpha_k = \frac{1}{\sqrt{2\pi}} \int_0^{2\pi} f(t) \, e^{-ikt} \, dt.$$

   Use this result to prove that $L_2[0, 2\pi]$ is isomorphic to l_2.

7. If $f \in L_2^*[0, 2\pi]$, prove that there exists $y \in L_2[0, 2\pi]$ such that

   $$f(x) = \int_0^{2\pi} x(t) \, \overline{y(t)} \, dt$$

   for all $x \in L_2[0, 2\pi]$.

3. The dual space of a Hilbert space

If H is a Hilbert space then one may immediately write down examples of bounded linear functionals on H. For f, defined by $f(x) = (x, y)$, y fixed, is linear, by definition of inner product. By Schwarz' inequality (see section 1) we also have $|f(x)| \leqslant \|x\| \, \|y\|$ for all $x \in H$, so that $\|f\| \leqslant \|y\|$. In fact $\|f\| = \|y\|$, as we see on putting $x = y$.

It is a remarkable fact that the inner products are the only bounded linear functionals on H. This result is known as the Riesz–Fréchet (or Riesz) representation theorem. It is proved below in theorem 13.

Now we give some preliminary theorems.

Theorem 10. *Let $G \neq H$ be a closed subspace of H and write $d = d(h, G)$, the distance of h from G, where $h \in H \sim G$. Then there exists $g \in G$ such that $\|h - g\| = d$, i.e. there is a point of G nearest to h.*

Proof. There exists $(x_n) \in G$ such that $\|h - x_n\| \to d$ $(n \to \infty)$. It is enough to show that (x_n) is Cauchy. Then, since G is complete (being closed) we have $x_n \to g$ $(n \to \infty)$, whence $\|h - g\| = d$. Now

$$\|x_n - x_m\|^2 = \|(h - x_m) + (x_n - h)\|^2$$
$$= 2(\|h - x_m\|^2 + \|x_n - h\|^2) - \|2h - (x_n + x_m)\|^2$$
$$= 2(\|h - x_m\|^2 + \|x_n - h\|^2) - 4\left\|h - \frac{x_n + x_m}{2}\right\|^2$$
$$\to 2(d^2 + d^2) - 4d^2 = 0 \quad (n, m \to \infty).$$

In the above we have used the so-called parallelogram law:

$$\|x + y\|^2 + \|x - y\|^2 = 2(\|x\|^2 + \|y\|^2),$$

and also the fact that

$$d \leqslant \left\|h - \frac{x_n + x_m}{2}\right\| \leqslant \tfrac{1}{2}\|h - x_n\| + \tfrac{1}{2}\|h - x_m\|.$$

We have thus shown that (x_n) is a Cauchy sequence, so the theorem is now proved.

Theorem 11. *Take the same hypothesis as in theorem 10. Then $(h - g) \perp G$, i.e. $(h - g, x) = 0$, for all $x \in G$.*

Proof. Suppose there exists $x \in G$ such that $\lambda = (h - g, x) \neq 0$. Then $x \neq \theta$ so we may define

$$k = g + \lambda \|x\|^{-2} x \in G.$$

Hence $\|h - k\|^2 = \|h - g\|^2 - |\lambda|^2 \|x\|^{-2} < \|h - g\|^2$, contrary to the fact that $\|h - g\| = d$.

Orthogonal sum

If M_1, M_2 are subspaces of an inner product space, then $M_1 + M_2$ is a subspace. When $M_1 \perp M_2$, i.e. $(m_1, m_2) = 0$ for every $m_1 \in M_1$, $m_2 \in M_2$, we write $M_1 \oplus M_2$ for $M_1 + M_2$, and call $M_1 \oplus M_2$ the orthogonal sum of M_1 and M_2.

Theorem 12. *Let G be a closed subspace of H. Write*

$$G^\perp = \{x \in H \,|\, x \perp G\}.$$

Then $H = G \oplus G^\perp$.

Proof. Let $h \in H$. If $h \in G$ it is clear that $h \in G + G^\perp$. Suppose that $h \in H \sim G$. By theorem 11, $h - g \in G^\perp$, whence $h = g + (h - g) \in G + G^\perp$. Hence $H = G + G^\perp$. Obviously $G \perp G^\perp$ and so $H = G \oplus G^\perp$.

Theorem 13 (Riesz–Fréchet). *Let $f \in H^*$, the continuous dual space of a Hilbert space H. Then there exists a unique $y \in H$ such that $f(x) = (x, y)$, for all $x \in H$, with $\|f\| = \|y\|$.*

Proof. Write $G = \mathrm{Ker}\,(f)$. Then G is a closed subspace of H. If $G = H$ then $f \equiv 0$, so we take $y = \theta$. Suppose $G \neq H$. By theorem 11, there exists a non-zero $z \perp G$. Consider

$$S = \{zf(x) - xf(z) \,|\, x \in H\}.$$

Then $S \subset G$, and since $z \perp S$ we have, for every x,

$$(zf(x) - xf(z), z) = 0,$$

$$f(x) \|z\|^2 = (x, z) f(z).$$

Hence $f(x) = (x, y)$, for every x, where $y = z \cdot \overline{f(z)} / \|z\|^2$. If $(x, y) = (x, y')$ for all $x \in H$ then $(x, y - y') = 0$ and so $(y - y', y - y') = 0$, whence $y = y'$. Since $f(y) = \|y\|^2$ and $|(x, y)| \leqslant \|x\| \|y\|$, we have $\|f\| = \|y\|$. This proves the theorem.

Exercises 3

1. Prove that the element g in theorem 10 is unique.
2. Suppose M_1, M_2 are subspaces of an inner product space. Show that $M_1 \perp M_2$ if and only if $\|m_1 + m_2\|^2 = \|m_1\|^2 + \|m_2\|^2$ whenever $m_1 \in M_1$, $m_2 \in M_2$.

3. Let S be a subset of an inner product space X and write $S = \{x \in X \mid x \perp S\}$. Prove that

   (i) $S^\perp$ is a closed subspace of X.

   (ii) $S \subset (S^\perp)^\perp \equiv S^{\perp\perp}$.

   (iii) $(S^{\perp\perp})^\perp = S^\perp$.

4. Let G be a closed subspace of H. Use $H = G \oplus G^\perp$ to show that $G^{\perp\perp} = G$.

5. Let X be a finite dimensional inner product space and Y a normed space. Prove that, in the notation of section 2, chapter 4, $B(X, Y) = L(X, Y)$.

6. By the Riesz–Fréchet theorem there is a natural map $T : H^* \to H$, given by $T(f) = y$. Prove that T is an additive surjective isometry, which is conjugate homogeneous, i.e. $T(\lambda f) = \overline{\lambda} T(f)$.

7. Let H be a Hilbert space. Prove that H^* is a Hilbert space under the inner product $(f_1, f_2) = (y_2, y_1)$, where for $k = 1, 2, f_k(x) = (x, y_k)$ for all $x \in X$.

7

MATRIX TRANSFORMATIONS IN SEQUENCE SPACES

1. Matrix and linear transformations

In this chapter we present some of the theory of matrix transformations and summability. The matrices we are concerned with are infinite, not finite, matrices and the convergence problems that arise make the subject part of analysis rather than of algebra.

Interest in general matrix transformation theory was, to some extent, stimulated by special results in summability theory which were obtained by Cesàro, Borel and others, at the turn of the century. It was however the celebrated German mathematician O. Toeplitz (1881–1940) who, in 1911, brought the methods of linear space theory to bear on problems connected with matrix transformations on sequence spaces. Toeplitz characterized all those infinite matrices $A = (a_{nk})$, $n, k = 1, 2, 3, \ldots$, which map the space c into itself, leaving the limit of each convergent sequence invariant. To be explicit, he gave the necessary and sufficient conditions on A for

$$y_n = \sum_{k=1}^{\infty} a_{nk} x_k$$

to converge for each n and to tend to l as $n \to \infty$, whenever $x_k \to l$ ($k \to \infty$). These famous 'Toeplitz conditions' will be given shortly—they come out quite quickly as an application of the Banach–Steinhaus theorem of section 3, chapter 4. Of course Toeplitz did not have this theorem, which dates from the 1920s, at his disposal. He used the methods of classical analysis for the difficult part of the theorem, employing a somewhat complicated *reductio ad absurdum* argument. However, his proof is quite instructive and the interested reader is referred to a readily accessible proof, see Hardy (1949).

We shall see that the Banach–Steinhaus theorem and related results are especially suited to dealing with many problems in the theory of matrix transformations and summability. With the aid of this theorem much of the theory becomes accessible to those who would normally have neither the time nor the propensity to follow the usual 'classical' approach. One must however keep things in perspective, for although the functional analytic method often smooths the path of *proof*, it must be remembered that many of the actual *results* were given to us by the classical analysts. Whilst on this subject we should perhaps inform the reader, who might be uninitiated, of the so-called schools of 'hard' and 'soft' analysis. When functional analysis was growing up, some famous classical analysts such as G. H. Hardy (1877–1947) are said to have referred to it as soft analysis. If they said this at all it would seem that they meant that it was largely concerned with existence rather than constructive proofs. For example, in chapter 4 the Banach–Steinhaus theorem yielded the existence of a continuous function whose Fourier series diverged at a point. It was in the nature of the proof that the explicit form of this function could not be given. On aesthetic grounds it would surely be agreed that this was something of a deficiency. However, even on the basis of the little that has been given in this introductory volume of ours, it is hoped that the reader will see functional analysis as worthwhile and beautiful *per se*. To my mind there is no hard or soft analysis—there is just analysis.

One may ask why we should study such special transformations as those given by matrices—why not general linear operators? The answer is that, in many cases, the most general linear operator on one sequence space into another is actually given by a matrix. To make this precise let us consider the space c_0 of null sequences. Suppose $A = (a_{nk})$, $n, k = 1, 2, \ldots$, is an infinite matrix and $x = (x_k) \in c_0$. Let us apply A to x:

$$Ax = \begin{pmatrix} a_{11} & a_{12} & \ldots \\ a_{21} & a_{22} & \ldots \\ \ldots & \ldots & \ldots \end{pmatrix} \begin{pmatrix} x_1 \\ x_2 \\ \vdots \end{pmatrix}$$

obtaining by the usual method of matrix multiplication (applied *formally* in our case, since we have infinite matrices),

$$Ax = \begin{pmatrix} a_{11}x_1 + a_{12}x_2 + & \ldots \\ a_{21}x_1 + a_{22}x_2 + & \ldots \\ \ldots & \ldots & \ldots \end{pmatrix}.$$

Hence, in a *formal* way, we map the sequence x into the sequence Ax where $(Ax)_n \equiv A_n(x)$ is given by

$$A_n(x) = \sum_{k=1}^{\infty} a_{nk}x_k,$$

provided the series converges for each n. When does $A : c_0 \to c_0$? Obvious sufficient conditions for this are now given.

Theorem 1. *Let $a_{nk} \to 0$ ($n \to \infty$, k fixed) and suppose*

$$M = \sup_n \Sigma |a_{nk}| < \infty,$$

where here and in future sums are over k, for each n. Then A defines a bounded linear operator on c_0 into itself and $\|A\| = M$.

Proof. Let $x \in c_0$. Then $Ax \in c_0$, i.e. $A_n(x) \to 0$ ($n \to \infty$). For we have that the series $\Sigma a_{nk} x_k$ is absolutely convergent for each n, and further, for any $m \geqslant 1$,

$$|A_n(x)| \leqslant \sum_{k=1}^{m} |a_{nk} x_k| + \sum_{k=m+1}^{\infty} |a_{nk} x_k|$$

$$\leqslant \|x\| \sum_{k=1}^{m} |a_{nk}| + \max_{k \geqslant m+1} |x_k| \cdot M.$$

Now take m so large that $\max \{|x_k| \, | k \geqslant m+1\} < \epsilon$ and then take n so large that $\sum_{1}^{m} |a_{nk}| < \epsilon$ (possible since $a_{nk} \to 0$ as $n \to \infty$, k fixed). Hence we have shown that $A : c_0 \to c_0$. It is clear that A is linear, e.g.

$$A(\lambda x) = (\Sigma a_{nk} \lambda x_k)_{n \in N} = \lambda(\Sigma a_{nk} x_k)_{n \in N} = \lambda A(x).$$

Finally, A is bounded:

$$\|A(x)\| = \sup_n |\Sigma a_{nk} x_k| \leqslant \|x\| \sup_n \Sigma |a_{nk}| = M\|x\|,$$

for every $x \in c_0$. It follows that $\|A\| \leqslant M$ and so we must show that $\|A\| = M$, which will complete the proof. Now there exists $m = m(\epsilon)$ such that $\Sigma |a_{mk}| > M - \epsilon/2$, and since $\Sigma |a_{mk}| < \infty$ there exists $p = p(\epsilon)$ such that

$$\sum_{k>p} |a_{mk}| < \epsilon/2.$$

Now define $x \in c_0$ by $x_k = \text{sgn}\, a_{mk}$, $1 \leqslant k \leqslant p$, $x_k = 0$, $k > p$. Then $\|x\| = 1$ and so

$$\|A(x)\|/\|x\| = \sup_n |A_n(x)| \geqslant |A_m(x)| > M - \epsilon.$$

It follows that $M = \sup \{\|A(x)\|/\|x\| \, | x \neq \theta\} = \|A\|$.

Theorem 1 shows that a certain type of matrix defines a bounded linear operator on c_0 into itself. Now we show the converse, viz. if $A \in B(c_0, c_0)$ then there exists (a_{nk}) such that

$$(Ax)_n = \Sigma a_{nk} x_k \quad \text{for every} \quad x \in c_0, \tag{1}$$

where the matrix (a_{nk}) must satisfy the conditions of theorem 1.

Theorem 2. *Let A be any bounded linear operator on c_0 into itself. Then A determines a matrix (a_{nk}) such that* (1) *holds and such that*

$$\|A\| = \sup_n \Sigma |a_{nk}| < \infty; \quad a_{nk} \to 0 \quad (n \to \infty, \ k \text{ fixed}).$$

Proof. Every $x \in c_0$ may be written as $x = \Sigma x_k e_k$, where $e_1 = (1, 0, 0, \ldots)$, $e_2 = (0, 1, 0, 0, \ldots) \ldots$, since (e_k) is a basis in c_0. Linearity and continuity of A yield

$$Ax = \Sigma x_k A e_k = \Sigma x_k (a_k^{(n)})_{n \in N},$$

where $A e_k$ is some sequence $(a_k^{(1)}, a_k^{(2)}, \ldots) \in c_0$, $k = 1, 2, \ldots$. Hence, since $|x_n| \leqslant \|x\|$ for every n, whenever $x \in c_0$, we obtain

$$(Ax)_n = \Sigma a_k^{(n)} x_k \quad (n = 1, 2, \ldots),$$

which we shall write, in other notation, as

$$A_n(x) = \Sigma a_{nk} x_k. \tag{2}$$

This proves (1). By our hypothesis that $Ax \in c_0$ whenever $x \in c_0$, we deduce that $A e_k \in c_0$, $k = 1, 2, \ldots$, which implies $a_{nk} \to 0$ $(n \to \infty)$, $k = 1, 2, \ldots$. It remains to show that $\|A\| = \sup_n \Sigma |a_{nk}|$. By theorem 1 it is enough to show that there exists H such that $\Sigma |a_{nk}| \leqslant H$ for all n. Now for each n, $\quad |A_n(x)| \leqslant \|A(x)\| \leqslant \|A\| \|x\|,$

so that A_n is a bounded (and obviously linear) functional on c_0. Thus we have a sequence $(A_n) \in c_0^*$ such that $\lim_n A_n(x) = 0$ on c_0. By the Banach–Steinhaus theorem it follows that the sequence $(\|A_n\|)$ of norms is bounded, i.e. $\quad \|A_n\| \leqslant H \quad$ for all n

and some constant H. But $\|A_n\|$ for A_n given by (2) is, by the proof of theorem 8, chapter 4 (see also the table of dual spaces),

$$\|A_n\| = \Sigma |a_{nk}|.$$

Hence we have shown that $M = \sup_n \Sigma |a_{nk}| < \infty$ and the proof of theorem 1 now gives $\|A\| = M$.

Using methods similar to that of theorem 2 is can be shown that for certain sequence spaces X, Y, every $A \in B(X, Y)$ is given by a matrix. Some of these spaces are now listed: (c_0, c_0), (c_0, c), (c_0, l_1), (c, c_0), (c, c), (c, l_1), (l_p, c_0), (l_p, c), (l_p, l_1), (l_p, l_s), where $1 \leqslant p < \infty$, $1 \leqslant s < \infty$. A notable exception is l_∞, which never occurs in the first entry in the bracket. This is due to the fact that not every element of l_∞^* can be expressed in the form of $\Sigma a_k x_k$. Since (e_k) is not a basis for c, we remark that slightly more care is needed in describing mappings from c.

Although, for example, $A \in B(l_2, l_2)$ can be represented as a matrix, it is not known at the present time, what the explicit necessary and sufficient conditions on $A = (a_{nk})$ are, to ensure that $A \in B(l_2, l_2)$. Sufficient conditions have been given by Schur (see Taylor, 1958). The proof Taylor gives depends on a knowledge of symmetric operators and eigenvalues. Later, we shall give a generalization of Schur's theorem the proof of which is completely elementary and uses only Hölder's inequality.

In view of what has been observed above we shall in future consider only matrix transformations on sequence spaces X, in the form

$$A_n(x) = \Sigma a_{nk} x_k.$$

By (X, Y) we shall denote the set of all matrices A which map X into Y. By $(X, Y; P)$ we denote that subset of (X, Y) for which limits or sums are *preserved*. For example, $A \in (c, c; P)$ means that $A_n(x)$ exists for each n whenever $x \in c$ and that $A_n(x) \to l(n \to \infty)$ whenever $x_k \to l$ $(k \to \infty)$.

Now we give some basic results.

Theorem 3 (Silverman–Toeplitz). $A \in (c, c; P)$ *if and only if*

(i) $\sup_n \Sigma |a_{nk}| < \infty$;

(ii) $a_{nk} \to 0 \ (n \to \infty, k \ fixed)$;

(iii) $\Sigma a_{nk} \to 1 \ (n \to \infty)$.

Proof. We remark that Silverman proved the sufficiency of the conditions, which is easy. The 'hard' part is the necessity of (i), which was Toeplitz' contribution. In future, a matrix satisfying (i)–(iii) will be called a Toeplitz matrix (some call it regular—a somewhat over-worked term).

The sufficiency of (i)–(iii) is established by writing

$$\Sigma a_{nk} x_k = \Sigma a_{nk}(x_k - l) + l \Sigma a_{nk}. \tag{3}$$

By (iii), the second term of (3) tends to l as $n \to \infty$. The first term tends to zero. To see this we use (i) and (ii), replacing x_k in theorem 1 by $x_k - l$.

The necessity of (ii) and (iii) is trivial—we just put $x = e_k \in c_0$, $k = 1, 2, \ldots$ for (ii) and $x = e = (1, 1, 1, \ldots) \in c$ for (iii). The necessity of (i) has essentially been dealt with in the proof of theorem 2. For completeness we indicate the argument. We are supposing that

$A_n(x) = \Sigma a_{nk} x_k$ exists for each n and tends to l whenever $x_k \to l$. The existence for each n and every $x \in c$ implies that $\Sigma |a_{nk}| < \infty$, each n, and

$$\|A_n\| = \Sigma |a_{nk}|. \tag{4}$$

This follows from the Banach–Steinhaus theorem or from the following elementary argument: eliminate n from the notation since it is fixed and suppose $\Sigma a_k x_k$ converges for all $x \in c$. Then $\Sigma |a_k| < \infty$; otherwise $\Sigma |a_k| = \infty$ and so there is a sequence of integers $n_1 < n_2 < n_3 < \cdots$ such that

$$\sum_i = \sum_{k=n_i+1}^{n_{i+1}} |a_k| > i.$$

Define $x \in c_0$ by $x_k = (\operatorname{sgn} a_k)/i$ for $n_i < k \leqslant n_{i+1}$ and $x_k = 0$ for $1 \leqslant k \leqslant n_1$. Then $\Sigma a_k x_k = \Sigma_1 + \Sigma_2/2 + \Sigma_3/3 + \cdots > 1 + 1 + \cdots$, so that $\Sigma a_k x_k$ diverges, contrary to supposition. Hence $\Sigma |a_k| < \infty$ and so $\Sigma a_k x_k$ defines an element of c^* since $|\Sigma a_k x_k| \leqslant \Sigma |a_k| \cdot \|x\|$ on c. Also the norm of this functional is $\Sigma |a_k|$ (see the table of dual spaces). Thus we have proved (4). Finally, we have $\limsup_n |A_n(x)| < \infty$ on c, by our hypothesis, whence the Banach–Steinhaus theorem yields

$$\sup_n \Sigma |a_{nk}| = \sup_n \|A_n\| < \infty,$$

which is (i).

There is a slight generalization of the Toeplitz theorem which concerns (c, c).

Theorem 4 (Kojima–Schur). $A \in (c, c)$ *if and only if*

   (i) $\sup_n \Sigma |a_{nk}| < \infty$;

   (ii) *for each p, there exists* $\lim_n \sum_{k=p}^{\infty} a_{nk} = a_p$.

Proof. The necessity of (i) is proved exactly as in theorem 3. The necessity of (ii) is obtained by taking $x = (0, 0, \ldots, 1, 1, 1, \ldots) \in c$, the first 1 being in the pth place.

For the sufficiency we write, when $x_k \to l$,

$$\Sigma a_{nk} x_k = \Sigma a_{nk}(x_k - l) + l\Sigma a_{nk} = s_n + t_n.$$

Then $t_n \to la_1$, by (ii), and we shall show that $s_n \to \Sigma b_k(x_k - l)$, where $b_k = \lim_n a_{nk} = a_k - a_{k+1}$, each k. Clearly

$$\Sigma |b_k| \leqslant \sup_n \Sigma |a_{nk}| < \infty,$$

and so $s_n - \Sigma b_k(x_k - l)$ is equal to

$$v_n = \Sigma(a_{nk} - b_k)(x_k - l).$$

The argument of theorem 1, with x_k replaced by $x_k - l$ and a_{nk} replaced by $a_{nk} - b_k$, shows that $v_n \to 0$ $(n \to \infty)$. Hence we have

$$\Sigma a_{nk} x_k \to l . \lim_n \Sigma a_{nk} + \Sigma (\lim_n a_{nk}) (x_k - l), \tag{5}$$

whenever $x_k \to l$.

There is quite an important object associated with (5):

Characteristic of a conservative matrix

Let $A \in (c, c)$. Then A is called a conservative (or convergence preserving) matrix and

$$\chi(A) = \lim_n \Sigma a_{nk} - \Sigma (\lim_n a_{nk})$$

is called the characteristic of A. The numbers $\lim_n a_{nk}$, $k = 1, 2, \ldots$ and $\lim_n \Sigma a_{nk}$ are referred to as the characteristic numbers of A.

By theorem 3 we see that the characteristic of a Toeplitz matrix is equal to 1. In general we define

Co-regular and co-null matrices

Let $A \in (c, c)$. Then A is co-regular if and only if $\chi(A) \neq 0$, and A is co-null otherwise.

Thus the Toeplitz matrices form a subset of the co-regular matrices, which in turn form a subset of the conservative matrices. Some properties of the characteristic will be found in the exercises. Also, we shall make use of it later in the chapter.

Next we give a characterization of the elements of (l_1, l_p), for $1 \leqslant p \leqslant \infty$. The case $p = 1$ was proved by K. Knopp and G. G. Lorentz in 1949.

Theorem 5. $A \in (l_1, l_p)$ if and only if

(a) $M = \sup_k \Sigma_n |a_{nk}|^p < \infty$, for the case $1 \leqslant p < \infty$,

(b) $\sup_{n,k} |a_{nk}| < \infty$, for the case $p = \infty$.

Proof. We prove (a), leaving (b) which has a similar proof, as an exercise. The sufficiency is just an application of Minkowski's inequality. For, if $x \in l_1$, then we want to show $(A_n(x)) \in l_p$:

$$(\Sigma_n |\Sigma_k a_{nk} x_k|^p)^{1/p} \leqslant \Sigma_k (\Sigma_n |a_{nk} x_k|^p)^{1/p}$$
$$= \Sigma_k |x_k| (\Sigma_n |a_{nk}|^p)^{1/p}$$
$$\leqslant \|x\| M^{1/p} < \infty.$$

The inversion of the summations over n and k is justified by absolute convergence. This deals with the sufficiency.

Now suppose $A \in (l_1, l_p)$ so that

$$\Sigma_i |A_i(x)|^p < \infty \quad \text{on} \quad l_1,$$

where $A_i(x) = \Sigma_k a_{ik} x_k$. Since $\Sigma_k a_{ik} x_k$ converges for each i, whenever $x \in l_1$, it follows by the Banach–Steinhaus theorem, or by a simple argument (like that leading to (4), theorem 3, above) that $\sup_k |a_{ik}| < \infty$, each i. Hence A_i defines an element of l_1^*, for each i. Now define

$$q_n(x) = \left(\sum_{i=1}^{n} |A_i(x)|^p \right)^{1/p} \quad (n = 1, 2, \ldots).$$

By Minkowski's inequality we see that each q_n is subadditive. Thus, since $q_n(\lambda x) = |\lambda| q_n(x)$ we have that each q_n is a seminorm on l_1. Moreover, the fact that each A_i is a bounded linear functional on l_1 implies that each q_n is bounded on l_1. Hence we have a sequence (q_n) of continuous seminorms on l_1 such that

$$\sup_n q_n(x) = \left(\sum_{i=1}^{\infty} |A_i(x)|^p \right)^{1/p} < \infty$$

for each $x \in l_1$. Applying theorem 11 of chapter 4 we obtain a constant H such that

$$\sum_{i=1}^{\infty} |A_i(x)|^p \leqslant H^p \|x\|^p, \quad \text{on} \quad l_1.$$

Putting $x = e_k$, $k = 1, 2, \ldots$, we get (a).

The next theorem, which was proved by Schur in 1921, is distinguished from the previous theorems both by the fact that the condition on the matrix is of a rather different nature and that the proof is 'classical'. No functional analytic proof of this theorem seems to be known—certainly the Banach–Steinhaus theorem, so useful in many connections, appears to be unsuited for the purpose.

To help the proof of Schur's theorem we give a simple lemma.

Lemma. *If $\Sigma |b_{nk}| < \infty$ for each n and $\Sigma |b_{nk}| \to 0$ $(n \to \infty)$ then $\Sigma |b_{nk}|$ is uniformly convergent in n.*

Proof. $\Sigma |b_{nk}| \to 0$ $(n \to \infty)$ *implies* $\sum_{k \geqslant 1} |b_{nk}| < \epsilon$, $n > N(\epsilon)$.

Since $\Sigma |b_{nk}| < \infty$, $1 \leqslant n \leqslant N(\epsilon)$, there exists $m = m(n, \epsilon)$ such that $\sum_{k \geqslant m} |b_{nk}| < \epsilon$. Hence there exists $M = M(\epsilon) \geqslant 1$ such that $\sum_{k \geqslant M} |b_{nk}| < \epsilon$ for all n, which means that $\Sigma |b_{nk}|$ is uniformly convergent in n.

We remark that the result of the lemma still holds with $|b_{nk}|$ replaced by non-negative b_{nk}.

Theorem 6 (Schur). $A \in (l_\infty, c)$ *if and only if* (i) $\Sigma_k |a_{nk}|$ *converges uniformly in* n; (ii) *there exists* $\lim_n a_{nk}$ (k *fixed*).

Proof. The sufficiency is trivial—$x \in l_\infty$ and (i) imply that $\Sigma a_{nk} x_k$ is absolutely and uniformly convergent in n, whence there exists $\lim_n \Sigma a_{nk} x_k = \Sigma a_k x_k$, where $a_k = \lim_n a_{nk}$.

The necessity of (ii) is obtained on taking $x = e_k \in l_\infty$ for $k = 1, 2, \ldots$. Now consider the necessity of (i). Since $(\Sigma a_{nk} x_k)$ converges, whenever $x \in l_\infty$, it converges whenever $x \in c$ and so, by the Kojima–Schur theorem, we must have $\sup_n \Sigma |a_{nk}| < \infty$. Write $b_{nk} = a_{nk} - a_k$, where $a_k = \lim_n a_{nk}$. Since $\Sigma |a_k| < \infty$ we have that $(\Sigma b_{nk} x_k)$ converges, whenever $x \in l_\infty$. We are going to show that this implies

$$\Sigma |b_{nk}| \to 0 \quad (n \to \infty). \tag{6}$$

Then, by the lemma preceding the theorem, it will follow that $\Sigma |b_{nk}|$ converges uniformly in n, whence $\Sigma |a_{nk}| = \Sigma |b_{nk} + a_k|$ converges uniformly in n, which is (i).

Thus we must prove (6), which we now do by *reductio ad absurdum*. Suppose then that $\Sigma |b_{nk}| \not\to 0$ $(n \to \infty)$. It follows that $\Sigma |b_{mk}| \to c > 0$ as $m \to \infty$ through some subsequence of the positive integers (for $\limsup_n \Sigma |b_{nk}| = c$, say). Also we have $b_{mk} \to 0$ $(m \to \infty$, each k). Hence we may determine $m(1)$ such that

$$|\Sigma |b_{m(1), k}| - c| < c/10 \quad \text{and} \quad |b_{m(1), 1}| < c/10.$$

Since $\Sigma |b_{m(1), k}| < \infty$ we may choose $k(2) > 1$ such that

$$\sum_{k=k(2)+1}^{\infty} |b_{m(1), k}| < c/10.$$

It follows that

$$\left| \sum_{k=2}^{k(2)} |b_{m(1), k}| - c \right| < 3c/10.$$

To contract notation we are going to write

$$\sum_{k=p}^{q} |b_{mk}| = B(m, p, q).$$

Now we choose $m(2) > m(1)$ such that $|B(m(2), 1, \infty) - c| < c/10$ and $B(m(2), 1, k(2)) < c/10$. Then choose $k(3) > k(2)$ such that

$$B(m(2), k(3) + 1, \infty) < c/10.$$

It follows that $|B(m(2), k(2) + 1, k(3)) - c| < 3c/10.$

Continue in this way and find $m(1) < m(2) < ..., 1 = k(1) < k(2) < ...$ such that

$$
\left.\begin{array}{c}
B(m(r), 1, k(r)) < c/10, \\
B(m(r), k(r+1)+1, \infty) < c/10, \\
|B(m(r), k(r)+1, k(r+1)) - c| < 3c/10.
\end{array}\right\} \tag{7}
$$

Let us define $x \in l_\infty$, $\|x\| = 1$, by $x_k = 0$ $(k = 1)$,

$$
x_k = (-1)^r \operatorname{sgn} b_{m(r), k} \, (k(r) < k \leqslant k(r+1)) \quad (r = 1, 2, ...).
$$

Then write $\Sigma b_{m(r), k} x_k$ as $\Sigma_1 + \Sigma_2 + \Sigma_3$, where Σ_1 is over $1 \leqslant k \leqslant k(r)$, Σ_2 over $k(r) < k \leqslant k(r+1)$ and Σ_3 over $k > k(r+1)$. It follows immediately from (7) and the definition of x that

$$
|\Sigma b_{m(r), k} x_k - (-1)^r c| < c/2.
$$

Consequently, it is clear that the sequence $(B_n(x)) = (\Sigma b_{nk} x_k)$ is not a Cauchy sequence and so not convergent. Thus if (6) is false then $(B_n(x))$ is not convergent for all $x \in l_\infty$, contrary to hypothesis. This completes the proof of Schur's theorem.

Corollary. *Strong and weak convergence of sequences coincide in l_1.*

Proof. We need only show that weak implies strong convergence. Suppose that $y^{(n)} \to y$ (weakly) in l_1, i.e. $f(y^{(n)} - y) \to 0$ as $n \to \infty$, for every $f \in l_1^*$. Now $f \in l_1^*$ if and only if there exists $x \in l_\infty$ such that $f(z) = \Sigma z_k x_k$ for all $z \in l_1$. Hence

$$
f(y^{(n)} - y) = \Sigma(y_k^{(n)} - y_k) x_k \equiv \Sigma b_{nk} x_k \to 0 \quad (n \to \infty)
$$

for every $x \in l_\infty$. By the proof of Schur's theorem it follows that

$$
\Sigma |y_k^{(n)} - y_k| \to 0 \quad (n \to \infty),
$$

i.e.
$$
\|y^{(n)} - y\| \to 0 \quad (n \to \infty),
$$

so that $y^{(n)} \to y$ (strongly) in l_1.

Our next theorem on matrix transformations concerns the space w_p of strongly Cesàro summable sequences, i.e. w_p is the set of all sequences $x = (x_k)$ such that there exists a number l for which

$$
\frac{1}{n} \sum_{k=1}^{n} |x_k - l|^p \to 0 \quad (n \to \infty).
$$

Here p is a fixed positive number. The number l is unique when it exists. For certain p the space w_p is of interest in ergodic theory (see Halmos, 1956) and in the theory of Fourier series (see Zygmund, 1955).

Natural norms and p-norms were defined for w_p in question 12, exercises 1, chapter 4, and w_p was complete in these norms.

Our object is to give necessary and sufficient conditions for an infinite matrix A to map w_p into c. For technical reasons we shall take norms in w_p different from those mentioned above. These new norms are however equivalent to the old ones. Define

$$\|x\| = \sup_r (2^{-r}\Sigma_r|x_k|^p)^{1/p} \quad (p \geqslant 1),$$

$$\|x\| = \sup_r 2^{-r}\Sigma_r|x_k|^p \quad (0 < p < 1),$$

where $r \geqslant 0$ and Σ_r denotes a sum over the range $2^r \leqslant k < 2^{r+1}$. For economy the dependence of $\|x\|$ on p has not been indicated, but it should be borne in mind. If we write

$$\|x\|^1 = \sup \left(n^{-1}\sum_{k=1}^n |x_k|^p\right)^{1/p} \quad (p \geqslant 1),$$

$$\|x\|^1 = \sup n^{-1}\sum_{k=1}^n |x_k|^p \quad (0 < p < 1)$$

then it is easy to check that, for any $p > 0$,

$$2^{-1}\|x\| \leqslant \|x\|^1 \leqslant 2\|x\|,$$

so that $\|x\|$ and $\|x\|^1$ are equivalent. For example, since w_p is complete with $\|x\|^1$ it is also complete with $\|x\|$. Thus in the case $p \geqslant 1$, $(w_p, \|\cdot\|)$ is a Banach space, and in the case $0 < p < 1$ it is a complete p-normed space. With regard to notation we write, for any matrix $A = (a_{nk})$ and $1/p + 1/q = 1$, $p \geqslant 1$,

$$A_r^p(n) = (\Sigma_r|a_{nk}|^q)^{1/q}, \tag{8}$$

where the sum is taken over k, for each n, and with k satisfying $2^r \leqslant k < 2^{r+1}$. The case $p = 1$ of (8) is interpreted as

$$A_r^1(n) = \max_r |a_{nk}|,$$

where for each n the maximum is taken for k such that

$$2^r \leqslant k < 2^{r+1}.$$

Theorem 7. (a) Let $0 < p < 1$. Then $A \in (w_p, c)$ if and only if

(i) $a_{nk} \to a_k \quad (n \to \infty, k \text{ fixed})$,

(ii) $M(A) = \sup_n \sum_{r=0}^{\infty} 2^{r/p} A_r^1(n) < \infty$.

(b) *Let $p \geqslant 1$. Then $A \in (w_p, c)$ if and only if*

(i)   *as in (a) above,*

(ii) $\sup_n \sum\limits_{r=0}^{\infty} 2^{r/p} A_r^p(n) < \infty$,

(iii) $\sum\limits_{k=1}^{\infty} a_{nk} \to a$ $(n \to \infty)$.

Proof. We prove part (a) only. Part (b) may be proved in a similar manner. First consider the sufficiency. When (ii) holds the series defining $A_n(x) = \Sigma a_{nk} x_k$ is absolutely convergent for each n. For, by the inequality
$$|\Sigma b_k|^p \leqslant \Sigma |b_k|^p \quad (0 < p < 1)$$

(see chapter 1, section 4), we have

$$\sum_{k=1}^{\infty} |a_{nk} x_k| = \sum_{r=0}^{\infty} \Sigma_r |a_{nk} x_k|$$

$$\leqslant \sum_{r=0}^{\infty} (\Sigma_r |a_{nk} x_k|^p)^{1/p}$$

$$\leqslant \sum_{r=0}^{\infty} A_r^1(n) . 2^{r/p} \|x\|^{1/p}$$

$$\leqslant M(A) \|x\|^{1/p} < \infty,$$

whenever $x \in w_p$. We shall now show that (ii) implies

$$\sum_{k=1}^{\infty} |a_{nk}| \quad \text{is uniformly convergent in } n. \tag{9}$$

Then (9) together with (i) implies

$$\lim_n \Sigma a_{nk} = \Sigma a_k. \tag{10}$$

To prove (9) we write $1/p + 1/q = 1$, $0 < p < 1$. Then for any positive integer s and for any $m \geqslant 2^s$, we have, for all n,

$$\sum_{k=m}^{\infty} |a_{nk}| \leqslant \sum_{r=s}^{\infty} \Sigma_r |a_{nk}| \leqslant \sum_{r=s}^{\infty} A_r^1(n) \, 2^r$$

$$\leqslant M(A) \, 2^{s/q}.$$

Since $q < 0$ it follows that (9) holds.

Now take $x \in w_p$ and suppose

$$n^{-1} \sum_{k=1}^{n} |x_k - l|^p \to 0 \quad (n \to \infty).$$

Write $\Sigma a_{nk} x_k = \Sigma a_k x_k + l\Sigma(a_{nk} - a_k) + \Sigma(a_{nk} - a_k)(x_k - l),$ (11)

and note that (i) and (ii) imply

$$\sum_{r=0}^{\infty} 2^{r/p} \max_r |a_k| \leqslant M(A).$$

Hence $\Sigma |a_k x_k| < \infty$ and the last sum in (11) has limit zero as $n \to \infty$. Thus, by (10), we now have $\Sigma a_{nk} x_k \to \Sigma a_k x_k$ as $n \to \infty$, for every x in w_p. This proves the sufficiency when $0 < p < 1$.

The necessity of (i) is trivial, so we turn to (ii). Suppose then that $A_n(x) = \Sigma a_{nk} x_k$ exists for each $n \geqslant 1$ whenever $x \in w_p$. Then for each n and each $r \geqslant 0$, the functionals $f_{rn}(x) = \Sigma_r a_{nk} x_k$ are in the dual space w_p^*; they are trivially linear, and continuous since

$$|f_{rn}(x)| \leqslant A_r^1(n)\, 2^{r/p} \|x\|^{1/p}.$$

It follows from the corollary to theorem 11, section 3, chapter 4, that

for each n, $\lim_s \sum_{r=0}^{s} f_{rn}(x) = A_n(x)$ is in w_p^*, whence

$$|A_n(x)| \leqslant \|A_n\| \|x\|^{1/p}. \tag{12}$$

For each n we take any integer $s > 0$ and define $x \in w_p$ by $x_k = 0$ for $k \geqslant 2^{s+1}$, $x_{N(r)} = 2^{r/p} \operatorname{sgn} a_{n,N(r)}, x_k = 0$ ($k \neq N(r)$) for $0 \leqslant r \leqslant s$, where $N(r)$ is such that $|a_{n,N(r)}| = \max_r |a_{nk}|$. By (12) we get

$$\sum_{r=0}^{s} 2^{r/p} A_r^1(n) \leqslant \|A_n\|,$$

whence for each n,

$$\sum_{r=0}^{\infty} 2^{r/p} A_r^1(n) \leqslant \|A_n\| < \infty. \tag{13}$$

Now the argument used in the sufficiency to prove that the series defining $A_n(x)$ was absolutely convergent gives

$$|A_n(x)| \leqslant \sum_{r=0}^{\infty} 2^{r/p} A_r^1(n) \|x\|^{1/p},$$

so that $$\|A_n\| \leqslant \sum_{r=0}^{\infty} 2^{r/p} A_r^1(n). \tag{14}$$

Together, (13) and (14) imply

$$\|A_n\| = \sum_{r=0}^{\infty} 2^{r/p} A_r^1(n).$$

Finally, by theorem 11, section 3, chapter 4, the existence of $\lim_n A_n(x)$ on w_p implies that

$$\sup_n \|A_n\| = \sup_n \sum_{r=0}^{\infty} 2^{r/p} A_r^1(n) < \infty,$$

which is (ii). The proof of (a) is now complete.

The following result on matrix transformations concerns the class $(\gamma, c; P)$. Thus we want necessary and sufficient conditions for a matrix B to be such that $\Sigma b_{nk} a_k \to s$ whenever $\Sigma a_k = s$.

Theorem 8. $B \in (\gamma, c; P)$ *if and only if*

$$\text{(i)} \quad b_{nk} \to 1 \quad (n \to \infty, \, k \text{ fixed}); \qquad \text{(ii)} \quad \sup_n \sum_{k=1}^{\infty} |\Delta b_{nk}| < \infty,$$

where $\Delta b_{nk} = b_{nk} - b_{n,k+1}$.

Proof. By Abel's partial summation, for each n, when (ii) holds and $\Sigma a_k = s$,
$$\Sigma b_{nk} a_k = s b_{n1} + \Sigma (s_k - s) \Delta b_{nk},$$

where $s_k = a_1 + a_2 + \ldots + a_k$. By (i) and (ii) it is clear that $(\Delta b_{nk}) \in (c_0, c_0)$, whence $\Sigma b_{nk} a_k \to s \ (n \to \infty)$.

The necessity of (i) is trivial and the necessity of (ii) may be made to depend on the necessity of the Toeplitz' theorem. The details are left to the reader. A matrix B in $(\gamma, c; P)$ is often called a *γ-matrix*.

In our final result on matrix transformations we give a simple proof of a generalization of a theorem of Schur. Schur's theorem was that, if $A \in (l_\infty, l_\infty) \cap (l_1, l_1)$ then $A \in (l_2, l_2)$.

Theorem 9. *Let* $1 < p < \infty$ *and suppose* $A \in (l_\infty, l_\infty) \cap (l_1, l_1)$. *Then* $A \in (l_p, l_p)$.

Proof. Using the methods of theorems 1–4 it is easy to show that $A \in (l_\infty, l_\infty)$ if and only if $\|A\|_\infty = \sup_n \Sigma_k |a_{nk}| < \infty$. Also, $A \in (l_1, l_1)$ if and only if $\|A\|_1 = \sup_k \Sigma_n |a_{nk}| < \infty$. Now

$$|\Sigma_k a_{nk} x_k| \leqslant \Sigma_k |a_{nk}|^{1/p} |a_{nk}|^{1/q} |x_k|,$$

where $1/p + 1/q = 1$. By Hölder's inequality,

$$\Sigma_k |a_{nk}|^{1/p} |a_{nk}|^{1/q} |x_k| \leqslant (\Sigma_k |a_{nk}| |x_k|^p)^{1/p} (\Sigma_k |a_{nk}|)^{1/q}$$

and so $\quad |\Sigma_k a_{nk} x_k|^p \leqslant (\Sigma_k |a_{nk}| |x_k|^p)(\Sigma_k |a_{nk}|)^{p/q}$.

Hence $\qquad \|Ax\|^p = \Sigma_n |\Sigma_k a_{nk} x_k|^p$

$$\leqslant \Sigma_n \Sigma_k |a_{nk}| |x_k|^p \|A\|_\infty^{p/q}$$

$$\leqslant \|A\|_\infty^{p/q} \Sigma_k |x_k|^p \Sigma_n |a_{nk}|$$

$$\leqslant \|A\|_\infty^{p/q} \|x\|^p \|A\|_1,$$

so that $\|Ax\| \leqslant \|A\|_\infty^{1/q} \|A\|_1^{1/p} \|x\|$, whenever $x \in l_p$.

Special matrices. We now give some examples of special Toeplitz and γ-matrices which are associated with particular authors. These matrices have been extensively used in the theory of summability, which we shall be examining in section 3. Before proceeding further we note that we may replace a_{nk} by $a_k(t)$ in most of our proofs and let $t \to \infty$ continuously. Thus, for example if

$$a_k(t) = e^{-k}t^k/k! \quad (k = 0, 1, 2, \ldots, t > 0),$$

then $a_k(t) \to 0$ $(t \to \infty, k$ fixed$)$ and

$$\sum_{k=0}^{\infty} |a_k(t)| = \sum_{k=0}^{\infty} a_k(t) = 1,$$

whence $\sup_t \sum |a_k(t)| < \infty$. We shall still call $A = (a_k(t))$ a Toeplitz matrix, although it is really a 'semicontinuous' matrix. Its essential property is that it maps convergent sequences into convergent functions, leaving the limit unchanged. In fact the matrix A above is called the Borel matrix, after the French mathematician E. Borel (1871–1956).

Borel matrix. The Toeplitz matrix defined by

$$a_k(t) = e^{-t} t^k/k! \quad (k = 0, 1, 2, \ldots, t > 0).$$

Arithmetic means. The Toeplitz matrix defined by

$$a_{nk} = 1/(n+1) \quad (0 \leqslant k \leqslant n), \qquad a_{nk} = 0 \quad (k > n).$$

(That $x_k \to x$ implies $\dfrac{1}{n+1} \sum_{k=0}^{n} x_k \to x$ was known to Cauchy.)

Cesàro means. For each $r > -1$ the (C, r) matrix is defined by

$$a_{nk} = A_{n-k}^{r-1}/A_n^r \quad (0 \leqslant k \leqslant n), \qquad a_{nk} = 0 \quad (k > n),$$

where $A_n^r = (r+1)(r+2) \ldots (r+n)/n!$ for $n \geqslant 1$, $A_0^r = 1$.
When $r \geqslant 0$, $A_n^{r-1} \geqslant 0$, and using the fact that

$$(1-z)^{-r-1} = \sum_{n=0}^{\infty} A_n^r z^n \quad (|z| < 1)$$

it is easy to show that

$$\sum_{k=0}^{n} A_{n-k}^s A_k^r = A_n^{s+r+1},$$

from which it follows that $\Sigma \left| a_{nk} \right| = \Sigma a_{nk} = 1$ for every n. Now it is known from elementary analysis that

$$\frac{A_n^r}{n^r} \to \frac{1}{\Gamma(r+1)} \qquad (n \to \infty), \tag{15}$$

where $\Gamma(x)$ denotes the gamma function. It follows that, for fixed k, $a_{nk} \to 0$ $(n \to \infty)$. The reader who is not familiar with (15) should turn to question 11, exercises 1.

Thus we have shown that each matrix (C, r), called the Cesàro matrix of order r, is a Toeplitz matrix when $r \geqslant 0$. When $-1 < r < 0$, (C, r) is not Toeplitz, but its properties may still of course be considered. The case $r = 0$ gives the unit matrix and the case $r = 1$ gives the arithmetic mean.

Euler–Knopp matrix. Define

$$a_{nk} = \binom{n}{k} r^k (1-r)^{n-k} \quad (0 \leqslant k \leqslant n), \qquad a_{nk} = 0 \quad (k > n),$$

where $\binom{n}{k} = n! / k! \, (n-k)!$. Then A is Toeplitz when $0 < r < 1$. We denote the Euler–Knopp matrix of order r by (E, r).

Abel matrix. This is the γ-matrix defined by $b_k(x) = x^k$, $0 < x < 1$, $k = 0, 1, \dots$. In this case we are taking the limit as $x \to 1-$ in $\Sigma b_k(x) a_k$. Abel's limit theorem (chapter 1, exercises 3) shows that B is a γ-matrix. Also, with the modification required for $x \to 1-$, one may employ theorem 8:

$$x^k \to 1 \; (x \to 1-, \, k \text{ fixed}); \qquad \sup_{0 < x < 1} \sum_{k=0}^{\infty} \left| \Delta x^k \right| = 1.$$

Riesz means. These matrices were introduced by Marcel Riesz (the brother of F. Riesz) and are important in the theory of Dirichlet's series (see G. H. Hardy and M. Riesz, 1915).

Let $0 \leqslant \lambda_0 < \lambda_1 < \lambda_2 < \dots < \lambda_n \to \infty$, $\mu > 0$ and define

$$b_k(t) = (1 - \lambda_k / t)^{\mu} \quad \text{for} \quad \lambda_k < t,$$

$$= 0 \qquad\qquad \text{for} \quad \lambda_k \geqslant t.$$

Then it is easy to check that B is a γ-matrix (here we let $t \to \infty$). We denote the Riesz mean of type $\lambda = (\lambda_n)$ and order μ by (R, λ, μ). It turns out that for $\lambda_n = n$ the matrix (R, n, μ) is 'equivalent' to the

Cesàro matrix (C, μ), i.e. x is summable (R, n, μ) to l if and only if x is summable (C, μ) to l (see p. 185).

There are many other special matrices which have found use in analysis. To mention a few, there are those associated with the names of Hausdorff, Hölder, Ingham, Lambert, Mittag-Leffler, Nörlund, Riemann and de la Vallee-Poussin. For the special properties of these matrices the reader is referred to Hardy's *Divergent Series*.

Exercises 1

1. Let X be a complex Banach space and let $c(X)$ be the set of all convergent sequences $x = (x_k)$, $x_k \in X$. Defining $\lambda x + \mu y = (\lambda x_k + \mu_k)$ and $\|x\| = \sup_n \|x_n\|$, prove that $c(X)$ is a Banach space.

   Suppose the matrix $A = (a_{nk})$ of complex numbers is such that $(\Sigma_k a_{nk} x_k)_{n \in N}$ converges whenever $x \in c(X)$. Prove that $A \in (c, c)$, i.e. A is conservative.

2. Prove the inclusions $(c, c; P) \subset (l_\infty, l_\infty)$, $(l_\infty, c) \subset (c, c)$, $(c, c; P) \subset (c_0, c_0)$ and show that $(c, c; P) \cap (l_\infty, c) = \varnothing$. What is the relation between (c_0, c_0) and (c, c)?

3. Prove that $\Sigma a_n x_n$ converges, whenever Σx_n has bounded partial sums, if and only if $a = (a_n) \in BV \cap c_0$. (Use Schur's theorem.)

4. Prove that $A \in (l_\infty, c_0)$ if and only if $\Sigma |a_{nk}| \to 0$ $(n \to \infty)$.

5. Let A be a non-negative Toeplitz matrix, i.e. $a_{nk} \geq 0$ for all n, k. If $A_n(x) = \Sigma a_{nk} x_k$, where x_k is real, prove that

$$\liminf x_n \leq \liminf A_n(x) \leq \limsup A_n(x) \leq \limsup x_n.$$

   Hence show that $x_n \to \infty$ implies $A_n(x) \to \infty$. Prove that $x_n \to \infty$ need not imply $A_n(x) \to \infty$ for positive conservative matrices.

6. A matrix A is called *totally regular* if $x_n \to s$ implies $A_n(x) \to s$ for all finite or infinite s. If $A_n(x) = 2x_n - x_{n+1}$ show that A is Toeplitz but not totally regular. Prove that the matrix

$$\begin{pmatrix} 1 & -2^{-1} & 2^{-2} & 2^{-3} & . & . \\ 0 & 1 & -2^{-2} & 2^{-3} & . & . \\ 0 & 0 & 1 & -2^{-3} & . & . \\ 0 & 0 & 0 & 1 & . & . \\ . & . & . & . & . & . \end{pmatrix}$$

   is totally regular.

7. Prove that $A \in (l_1, l_1; P)$, where $a_{nk} = k/n(n+1)$ for $1 \leq k \leq n$, $a_{nk} = 0$ for $k > n$.

8. Prove every matrix in (l_∞, c) is co-null.

9. Prove that $A, B \in (c, c)$ implies $A + B, AB \in (c, c)$, where AB denotes the usual matrix product:

$$(AB)_{nk} = \sum_{i=1}^{\infty} a_{ni} b_{ik}.$$

Show also that $\chi(A+B) = \chi(A) + \chi(B)$, $\chi(AB) = \chi(A)\chi(B)$ and $|\chi(A)| \leqslant \|A\| = \sup_n \Sigma |a_{nk}|$, whenever $A, B \in (c, c)$. Here $\chi(A)$ is the characteristic of the conservative matrix A.

10. $A \in (c, c)$ is called 'multiplicative m' if
$$\Sigma a_{nk} x_k \to m \lim x_k \quad (n \to \infty),$$
whenever $x \in c$. Find the characteristic of a matrix which is multiplicative m. When is such a matrix co-regular?

11. In connection with (C, r) means it is necessary to show that $\lim_n A_n^r/n^r$ exists and is non-zero (the limit is actually $1/\Gamma(r+1)$). Do this by considering $\log A_n^r$ and then expanding $\log(1 + r/k)$, $1 \leqslant k \leqslant n$, to second orders in $1/k$.

12. The series Σz^n converges if and only if $|z| < 1$. Prove that the Abel matrix 'sums' this series to $1/(1-z)$ for $|z| = 1$ but $z \neq 1$. By this we mean that there exists
$$\lim_{x \to 1-} \Sigma z^n x^n = 1/(1-z) \quad \text{if} \quad |z| = 1, \quad z \neq 1.$$
Thus $1 - 1 + 1 - \ldots = \frac{1}{2}$, in the sense of 'sum' just defined. More precisely we say that $1 - 1 + 1 - 1 + \ldots = \frac{1}{2}$ (Abel).

13. Prove that $\Sigma z^n = 1/(1-z)$ (Borel), for $Rl(z) < 1$. Show also that $\Sigma z^n = 1/(1-z)$ (Euler), for $|1 - r + rz| < 1$.

14. If A is a Toeplitz matrix and B is a γ-matrix prove that AB is a γ-matrix. Show that BA need not even exist.

15. Let $p_0 > 0$, $p_n \geqslant 0$ $(n \geqslant 1)$, $P_n = p_0 + \ldots + p_n$. Define $a_{nk} = p_{n-k}/P_n$ $(0 \leqslant k \leqslant n)$, $a_{nk} = 0$ $(k > n)$ and prove that A is Toeplitz if and only if $p_n/P_n \to 0$ $(n \to \infty)$. The matrix A defines the *Nörlund mean*.

16. Let A be the (C, r) matrix and write
$$t_n^r = \sum_{k=0}^{n} a_{nk} x_k.$$
For $r > -1, h > 0$ prove that
$$A_n^{r+h} t_n^{r+h} = \sum_{k=0}^{n} A_{n-k}^{h-1} A_k^r t_k^r$$
and hence show that $t_k^r \to s$ $(k \to \infty)$ implies $t_n^{r+h} \to s$ $(n \to \infty)$.

17. Prove part (b) of theorem 7, section 1, chapter 7.

18. Let $A = (a_{nk})$ be a complex Toeplitz matrix. Find whether $Rl(A)$ and $Im(A)$ are Toeplitz matrices. Here we mean $Rl(A) = (Rl(a_{nk}))$, etc.

2. Algebras of matrices

There are several ways of combining infinite matrices to give some kind of product matrix. The most natural product is perhaps the usual *matrix product*, which arises on iterating transformations. Thus, arguing formally, if $y = Bx = (\Sigma_k b_{ik} x_k)$, then
$$Ay = (\Sigma_i a_{ni} y_i) = (\Sigma_i a_{ni} \Sigma_k b_{ik} x_k) = (\Sigma_k x_k \Sigma_i a_{ni} b_{ik}),$$
so that
$$(AB)_{nk} = \Sigma_i a_{ni} b_{ik}. \tag{16}$$

Whenever we write the matrix product AB it will be understood that the (n, k) element of AB is defined by (16). Thus AB is well-defined only when the series in (16) converge for every n, k. Matrix multiplication is not commutative, as is easy to see, and it can happen that AB exists but BA does not (see the exercises). The matrix product makes the set of conservative matrices into a Banach algebra. Before we prove this we turn to other types of product.

If we regard a matrix as nothing but a double array of numbers then the simplest way of forming a product of matrices A, B is to multiply corresponding terms. The (n, k) element of the product being $a_{nk}b_{nk}$. This is the *term product* of two matrices—it seems to be of little use and we shall not introduce notation for it, though we shall occasionally mention it.

Another type of product, called the *convolution product* (or just convolution) is obtained if one transforms a power series Σz^k with matrices A, B and then multiplies the transformations according to Cauchy's rule:

$$(\Sigma_k a_{nk} z^k) \cdot (\Sigma_k b_{nk} z^k) = \left(\Sigma_k z^k \sum_{i=0}^{k} a_{ni} b_{n, k-i} \right).$$

Thus we define the convolution A_*B by

$$(A_*B)_{nk} = \sum_{i=0}^{k} a_{ni} b_{n, k-i}. \tag{17}$$

Unlike the matrix product, both the term product and the convolution are well-defined for all matrices, since no questions of convergence arise.

The only other product we shall consider is the *second convolution* $A_{**}B$, whose (n, k) element is defined as

$$\frac{1}{k+1} \sum_{i=0}^{k} a_{ni} b_{n, k-i}. \tag{18}$$

The second convolution product is suggested by transforming a power series Σz^k and then integrating with respect to z the Cauchy product of the transformations.

Our first result concerns the class (c, c) of conservative matrices.

Theorem 10. *The conservative matrices form a Banach algebra with identity under the matrix product.*

Proof. If we define $\|A\| = \sup_n \Sigma |a_{nk}|$ and $\lambda A + \mu B = (\lambda a_{nk} + \mu b_{nk})$ then it is clear that (c, c) is a normed linear space. Now let $A, B \in (c, c)$. Then
$$\|AB\| = \sup_n \Sigma_k |\Sigma_i a_{ni} b_{ik}|$$
$$\leqslant \sup_n \Sigma_i |a_{ni}| \Sigma_k |b_{ik}| \leqslant \|A\| \|B\| < \infty.$$

Write $C = AB$ and let $x \in c$. Then, by the absolute convergence of the double series, which we have just shown,
$$C_n(x) = \Sigma_k c_{nk} x_k = \Sigma_k (\Sigma_i a_{ni} b_{ik}) x_k$$
$$= \Sigma_i a_{ni} (\Sigma_k b_{ik} x_k) = A_n(Bx),$$

so that $C(x) = A(Bx)$. Since B is conservative, $Bx \in c$ and since A is conservative, $C(x) \in c$. Thus (c, c) is closed under the matrix product.

Next, we check the associative law, leaving the checking of the other algebraic laws as an exercise. Let A, B, C be in (c, c). Since we have just shown that $(AB)(x) = A(Bx)$ for each $x \in c$, it follows immediately that $[(AB)C](x) = [A(BC)](x)$ for each $x \in c$, whence, putting $x = e_k$ we get $(AB)C = A(BC)$.

The identity in the algebra is the unit matrix I, whose elements are 1 in the diagonal and zero elsewhere, i.e. $I = (\delta_{nk})$, with $\delta_{nn} = 1$ and $\delta_{nk} = 0$ $(n \neq k)$.

Finally, we show that (c, c) is a complete space by proving that every absolutely convergent series is convergent (see theorem 2, chapter 4). Let
$$A^{(r)} = (a_{nk}^{(r)}) \in (c, c) \quad (r = 1, 2, \ldots),$$

and suppose that $\Sigma \|A^{(r)}\| < \infty$. Let us write $A_n^{(r)} = \Sigma_k a_{nk}^{(r)}$, so that for each r, we have the existence of $\lim_n A_n^{(r)} = a^{(r)}$ and $\lim_n a_{nk}^{(r)} = a_k^{(r)}$ (k fixed). We shall show that $\Sigma A^{(r)}$ converges in the norm to $A = (\Sigma_r a_{nk}^{(r)}) \in (c, c)$. First, A exists, since for all n, k,
$$\Sigma_r |a_{nk}^{(r)}| \leqslant \Sigma_r \|A^{(r)}\| < \infty.$$

Also, for all n,
$$\Sigma_r \Sigma_k |a_{nk}^{(r)}| \leqslant \Sigma_r \|A^{(r)}\|, \tag{19}$$

whence
$$\Sigma_k |a_{nk}| = \Sigma_k |\Sigma_r a_{nk}^{(r)}| \leqslant \Sigma_r \|A^{(r)}\|,$$

the interchange of the order of summation being justified by absolute convergence. It now follows that $\|A\| \leqslant \Sigma_r \|A^{(r)}\|$. Again, by the absolute convergence expressed by (19), we have
$$\Sigma_k a_{nk} = \Sigma_r \Sigma_k a_{nk}^{(r)} = \Sigma_r A_n^{(r)}.$$

By hypothesis, $A_n^{(r)} \to a^{(r)}$ $(n \to \infty$, each $r)$ and by (19), the series $\Sigma_r A_n^{(r)}$ is uniformly convergent in n (Weierstrass M-test). It follows that

$$\Sigma_k a_{nk} \to \Sigma_r a^{(r)} \quad (n \to \infty).$$

That $$a_{nk} \to \Sigma_r a_k^{(r)} \quad (n \to \infty, \ k \text{ fixed})$$

may be shown in a similar way. Thus we have shown that $A \in (c, c)$. Finally, $\Sigma A^{(r)}$ converges in norm to A. For

$$\sum_{k=1}^{\infty} \left| \sum_{r=m+1}^{\infty} a_{nk}^{(r)} \right| \leqslant \sum_{r=m+1}^{\infty} \sum_{k=1}^{\infty} |a_{nk}^{(r)}| \leqslant \sum_{r=m+1}^{\infty} \|A^{(r)}\|$$

and so

$$\left\| \sum_{r=1}^{m} A^{(r)} - A \right\| = \left\| \left(\sum_{r=m+1}^{\infty} a_{nk}^{(r)} \right) \right\| \leqslant \sum_{r=m+1}^{\infty} \|A^{(r)}\| \to 0 \quad (m \to \infty).$$

This proves theorem 10.

Now let us consider the set $(c, c; P)$ of Toeplitz matrices as a subset of (c, c). It is obvious that $(c, c; P)$ is not a subspace, since $2\Sigma a_{nk} \to 2$ $(n \to \infty)$, whenever A is Toeplitz. The properties that $(c, c; P)$ does have are given in

Theorem 11. *As a subset of (c, c), the set $(c, c; P)$ is a closed convex semigroup with identity.*

Proof. Let A be in the closure of $(c, c; P)$. Then, with $\|A\| = \sup_n \Sigma |a_{nk}|$, there exists $A^{(m)} \in (c, c; P)$ such that $\|A^{(m)} - A\| < \epsilon/2$. Also,

$$|a_{nk}| \leqslant \|A^{(m)} - A\| + |a_{nk}^{(m)}|,$$

whence $a_{nk} \to 0$ $(n \to \infty, \ k$ fixed$)$, since $a_{nk}^{(m)} \to 0$ $(n \to \infty, \ k$ fixed$)$. Again,

$$|\Sigma_k a_{nk} - 1| \leqslant \|A^{(m)} - A\| + |\Sigma_k a_{nk}^{(m)} - 1|,$$

and so $\Sigma_k a_{nk} \to 1$ $(n \to \infty)$, whence by theorem 3, $A \in (c, c; P)$; thus $(c, c; P)$ is closed.

It is clear that $(c, c; P)$ has a stronger property than convexity. Obviously, $\lambda A + \mu B \in (c, c; P)$ whenever $\lambda + \mu = 1$ and $A, B \in (c, c; P)$. In fact it can be shown that $(\Sigma_r \lambda_r a_{nk}^{(r)}) \in (c, c; P)$ whenever $A^{(r)} \in (c, c; P)$ for $r = 1, 2, \ldots, \Sigma_r \lambda_r = 1$ and $\Sigma_r |\lambda_r| \|A^{(r)}\| < \infty$ (see exercise 4).

We are assuming the matrix product for the semigroup part of the theorem and of course the unit matrix is a Toeplitz matrix which is the identity. It remains to check closure under the product. Now if $C = AB$, the proof of theorem 10 gives $C_n(x) = A_n(Bx)$ for each $x \in c$. Hence $\lim_n C_n(x) = \lim_n A_n(Bx) = \lim_n B_n(x)$, since $A \in (c, c; P)$. Also,

MEO

$\lim_n B_n(x) = \lim_n x_n$, since $B \in (c, c; P)$. Thus AB is a Toeplitz matrix whenever A, B are Toeplitz matrices.

Of course, $(c, c; P)$ is not a group. For example, it is easy to check that the Toeplitz matrix

$$\begin{pmatrix} 0 & 0 & 0 & 0 & . & . & . \\ 0 & 1 & 0 & 0 & . & . & . \\ 0 & 0 & 1 & 0 & . & . & . \\ 0 & 0 & 0 & 1 & . & . & . \\ . & . & . & . & . & . & . \end{pmatrix}$$

has neither a left nor a right inverse. Even if a Toeplitz matrix has an inverse, this inverse may not be a Toeplitz matrix (see exercise 7).

Next we consider convolutions of matrices. By S we shall denote the set of all infinite matrices, with the usual co-ordinatewise addition and scalar multiplication. It is clear by (17) that S is a commutative algebra. There is also a convolution identity E, which is in fact the conservative matrix whose first column consists of 1's and which has zeros elsewhere:

$$e_{n0} = 1 \quad (n = 0, 1, \ldots); \qquad e_{nk} = 0 \quad (k > 0).$$

If we keep the usual norm structure in (c, c), so that (c, c) is a Banach space, but use convolution rather than matrix product, then (c, c) becomes a commutative Banach algebra with identity:

Theorem 12. *The set of conservative matrices is a commutative Banach convolution algebra with identity.*

Proof. In view of our remarks immediately above, we have only to show closure under convolution and the submultiplicative property of the norm. This is trivial—if the characteristic numbers of A, B, are a, a_k, b, b_k then by (17), with $C = A_* B$, we have

$$c_{n, k} \to a_0 b_k + \ldots + a_k b_0 \quad (n \to \infty, k \text{ fixed}).$$

Also, by the elementary theorem on the Cauchy product of absolutely convergent series, we have for each n,

$$\Sigma_k |c_{nk}| \leqslant \Sigma_k |a_{nk}| . \Sigma_k |b_{nk}|, \quad \Sigma_k c_{nk} = \Sigma_k a_{nk} . \Sigma_k b_{nk}.$$

Hence $\|C\| \leqslant \|A\| . \|B\|$ and $\Sigma_k c_{nk} \to ab \quad (n \to \infty)$,

so that C is seen to be conservative and the norm is submultiplicative. We may also note that the identity E has norm equal to 1.

Corollary. *As a subset of the convolution algebra* (c, c), *the Toeplitz matrices form a semigroup without identity.*

Proof. This is an easy exercise.

Finally, we make one or two remarks concerning the second convolution given by (18). Under the second convolution the set S of all matrices is a commutative groupoid‡ without identity. Closure and commutativity are obvious. If we suppose that an identity E exists, then $A_{**}E = A$ for all A. Putting $a_{nk} = 1$ we obtain $e_{nk} = 1$. Then, on putting $a_{nk} = k$ in $A_{**}E = A$ we obtain a contradiction. Hence there is no identity. S is not a semigroup, since the associative law fails. For example, if

$$A = B = \begin{pmatrix} 1 & 0 & 0 & \dots \\ 1 & 1 & 0 & \dots \\ 0 & 0 & 0 & \dots \\ \dots\dots\dots\dots \end{pmatrix}, \quad C = \begin{pmatrix} 1 & 0 & 0 & \dots \\ 0 & 1 & 0 & \dots \\ \dots\dots\dots\dots \end{pmatrix}$$

then the $(1, 1)$ element of $(A_{**}B)_{**}C$ is $1/2$ but the $(1, 1)$ element of $A_{**}(B_{**}C)$ is $1/4$.

It can be shown that the set (γ, c) is a Banach space which is a commutative non-associative algebra without identity, under the second convolution (see exercise 9). We shall prove here merely that the set of γ-matrices is a subgroupoid of S.

Theorem 13. $(\gamma, c; P)$ *is a subgroupoid of* S, *under the second convolution.*

Proof. That $(\gamma, c; P)$ has no identity and is not a semigroup may be shown as above. Now if $C = A_{**}B$, with A, B in $(\gamma, c; P)$, then by (18), since $a_{nk} \to 1, b_{nk} \to 1$ ($n \to \infty$, k fixed), we have $c_{nk} \to 1$ ($n \to \infty$, k fixed). From the easily checked identity (omitting n for simplicity),

$$(i+1) \sum_{j=0}^{i-1} a_j b_{i-1-j} - i \sum_{j=0}^{i} a_j b_{i-j}$$

$$= \sum_{j=1}^{i} j\{a_{i-j}(b_{j-1} - b_j) + b_{i-j}(a_{j-1} - a_j)\}$$

it follows that

$$\sum_{i=1}^{k} |c_{i-1} - c_i| \leqslant \sum_{i=1}^{k} \sum_{j=1}^{k} i^{-1}(i+1)^{-1} j\{|a_{i-j}||\Delta b_j| + |b_{i-j}||\Delta a_j|\},$$

† A groupoid is a pair $(G, .)$; G a nonempty set, and . a function on $G \times G$ into G. In our case, $G = S$ and . is $_{**}$.

where $\Delta a_j = a_{j-1} - a_j$. Since

$$|a_{nk}| \leqslant |a_{n0}| + |a_{n0} - a_{n1}| + \ldots + |a_{n,\,k-1} - a_{nk}|$$

and

$$a_{n0} \to 1 \quad (n \to \infty),$$

$$\sup_n \sum_{k=1}^{\infty} |a_{n,\,k-1} - a_{n,\,k}| = M(A) < \infty,$$

we have

$$\sum_{i=1}^{k} |\Delta c_i| \leqslant \{\sup |a_{n0}| + M(A)\}\, M(B) + \{\sup |b_{n0}| + M(B)\}\, M(A).$$

Recalling the fact that $c_i = c_{ni}$ we now see that

$$\sup_n \sum_{i=1}^{\infty} |\Delta c_{n,\,i}| < \infty.$$

Hence $C = A_{**}B$ is a γ-matrix.

Exercises 2

1. $A = (a_{nk})$ is called a *lower semimatrix* (or lower triangular matrix) if $a_{nk} = 0 \; (k > n)$. Prove that the set Δ of all conservative lower semi-matrices is an algebra and that every matrix in the radical of Δ has zero diagonal.

2. Let A be a Toeplitz matrix and B a conservative matrix. Prove that the characteristic numbers of AB are the same as those of B. Give an example in which the characteristic numbers of BA are different from those of B.

3. Prove that the set of co-regular matrices is an open subset of (c, c), but is not a subspace.

4. Let $A^{(r)} \in (c, c; p)$, $r = 1, 2, \ldots, \Sigma_r \lambda_r = 1$ and $\Sigma_r |\lambda_r| \, \|A^{(r)}\| < \infty$. Show that $(\Sigma_r \lambda_r a_{nk}^{(r)}) \in (c, c; p)$.

5. (i) Prove that $(\gamma, c; p)$ is a groupoid under the term product. Is it an algebra?

   (ii) Is (c, c) an algebra under the term product?

6. Is (l_∞, c) a closed subalgebra of (c, c), under the matrix product?

7. Show that the Toeplitz matrix of $(C, 1)$ means has an inverse which is not conservative.

8. Give an example of two conservative matrices A, B, which are not Toeplitz matrices, such that A_*B is Toeplitz.

9. Prove that $A \in (\gamma, c)$ if and only if $a_{nk} \to a_k \; (n \to \infty, \, k \text{ fixed})$,

$$M(A) = \sup_n \sum_{k \geqslant 1} |a_{n,\,k-1} - a_{n,\,k}| < \infty.$$

Define

$$\|A\| = \sup_n |a_{n0}| + 2M(A).$$

Show that this is a norm on (γ, c) and that (γ, c) is a Banach space which is a commutative non-associative algebra without identity, under the second convolution.

10. Let E be that subset of (γ, c) for which $\lim_k a_{nk} = 0$, $n = 0, 1, \ldots$. Prove that E is an ideal in (γ, c), under the second convolution. (The condition $\lim_n a_{nk} = 0$ is significant in that it is necessary for a matrix $A \in (\gamma, c)$ to be *stronger than convergence*. For the meaning of this last expression reference should be made to section 3 below.)

3. Summability

There are three main types of summability of sequences by infinite matrices—ordinary, absolute and strong. First we consider 'ordinary' summability, reserving absolute and strong to distinguish the other types.

Take any matrix $A = (a_{nk})$. Then a sequence $x = (x_k)$ is said to be *summable A to l*, written $x_k \to l(A)$, if and only if $A_n(x) = \Sigma_k a_{nk} x_k$ exists for each n and $A_n(x) \to l$ $(n \to \infty)$. For example, if I is the unit matrix, then $x_k \to l(I)$ means precisely that $x_k \to l$ $(k \to \infty)$, in the ordinary sense of convergence.

We denote by (A) the set of all sequences which are summable A. The set (A) is called the *summability field* of the matrix A. Thus, if $Ax = (A_n(x))$, then $(A) = \{x \,|\, Ax \in c\}$, where c is the set of convergent sequences. For example, $(I) = c$. The inclusion $c \subset (A)$ holds if and only if A is a conservative matrix. In particular, $c \subset (A)$ when A is a Toeplitz matrix, but here the set inclusion masks the more precise information that $x \in c$, $x_k \to l$, implies $A_n(x) \to l$, and not merely $Ax \in c$.

Historical material and detailed properties of special summability methods may be found in Hardy's excellent book *Divergent Series* (1949). We confine ourselves largely to more general questions. However, there is one point that we should first mention. Convergence of sequences and series, as we understand it today, was precisely defined by Cauchy. Many years before, Léonard Euler (1707–83), perhaps the greatest analyst of all time, was performing miraculous feats with series. He made great steps forward and discovered many relationships and beautiful formulae, as the reader will know from elementary analysis. Euler did not have a rigorous definition of convergence and he was quite prepared to put $x = 1$ in the formula $(1+x)^{-1} = 1 - x + x^2 - \ldots$ (which we know is valid only for $|x| < 1$) and obtained the curious result that $1/2 = 1 - 1 + 1 - \ldots$. Within the confines of our $(\epsilon, N(\epsilon))$ definition of convergence the formula just cited is forever meaningless—the right-hand side is a symbol for a certain limit (of the sequence $(1, 0, 1, 0, \ldots)$) and this limit does not exist, whence it cannot be $1/2$. However, the concept of summability does allow meaning to be given to the formula. If $x = (1, 0, 1, 0, \ldots)$ then it is easy to check that x is summable $(C, 1)$ to $1/2$, which we may write as $x_k \to 1/2(C, 1)$, or in series form, $1 - 1 + 1 - \ldots = 1/2(C, 1)$. It is also the case that $1 - 1 + 1 - \ldots = 1/2$ (Abel). For this we must show that $\Sigma a_k x^k \to 1/2 \ (x \to 1-)$, where $a_k = (-1)^k$. But, for $|x| < 1$, we have

$$\Sigma a_k x^k = 1 - x + \ldots = (1 + x)^{-1} \to 1/2 \quad \text{as} \quad x \to 1-.$$

Thus we have two particular matrices or summability methods which 'sum' the divergent series $1 - 1 + 1 - \ldots$ to the value $1/2$.

The above example is interesting but rather trivial. To mention but three things, summability has had success in the theory of Fourier's series, where Fejér's theorem on $(C, 1)$ summability was perhaps the first breakthrough after years of little advance. Incidentally, a corollary of Fejér's theorem is the Weierstrass approximation theorem (essentially saying that the polynomials on $[0, 1]$ are dense in $C[0, 1]$, with the supremum norm). Also, the Tauberian theory of Norbert Wiener, led him, using Lambert summability, to a proof of the prime number theorem. Thirdly, the multiplication of series is put in a satisfactory form by using Cesàro summability.

Now let us turn to some general considerations. If A is a Toeplitz matrix then $c \subset (A)$—more precisely A sums every convergent sequence to the same limit. It may happen, as we have already seen that the inclusion is strict. For example $x = (1, 0, 1, 0, \ldots) \in (C, 1)$ but $x \notin c$. Thus the $(C, 1)$ mean sums at least one divergent sequence. We say that $(C, 1)$ is *stronger than convergence*. Similarly for the general case of a matrix A. It is possible that a Toeplitz matrix A sums only convergent sequences, i.e. $(A) = c$. We then say that A is *equivalent to convergence*. Apart from the obvious case in which $(I) = c$ it does not seem to be known exactly what makes a Toeplitz matrix equivalent to convergence. A theorem which proves that $(A) = c$ is called a *Mercerian theorem*, after Mercer, who proved a significant theorem of this type. We may describe his result as follows. Let A be defined by $A_n(x) = \alpha x_n + (1 - \alpha) m_n$ where $\alpha > 0$ and m_n is the arithmetic mean of $x_1, x_2, \ldots, x_n$. Then A is Toeplitz, since $x_k \to l$ implies

$$A_n(x) \to \alpha l + (1 - \alpha) l = l$$

(actually without restriction on α). Mercer proved that, when $\alpha > 0$, $A_n(x) \to l$ implies $x_k \to l$. For a proof of this we refer to Hardy's *Divergent Series*, theorem 51.

A given Toeplitz matrix A may sum certain divergent sequences. It was first proved by Steinhaus that A cannot sum all bounded sequences. In fact he proved that there exists a sequence of 0's and 1's which is not A summable. We shall deduce Steinhaus' theorem from our results on matrix transformations. A proof of the original theorem may be found in Cooke's book, *Infinite Matrices and Sequence Spaces*, theorem 4.4 (III).

Theorem 14 (Steinhaus). *Given any Toeplitz matrix A there is always a bounded sequence which is not summable A.*

Proof. The characteristic $\chi(A) = 1$. Thus $A \notin (l_\infty, c)$, since matrices in (l_∞, c) have zero characteristic (see theorem 6, (i) and (ii)). But $A \in (l_\infty, c)$ if and only if A sums all bounded sequences, whence the result. We remark that 'Toeplitz' may be replaced by 'co-regular' in the hypothesis of the theorem, thus giving a slightly more general result.

Already we have remarked that a Toeplitz matrix may be equivalent to convergence. We now show that certain types of co-null matrices are always stronger than convergence. A matrix A is called *normal* if $a_{nk} = 0$ $(k > n)$ and $a_{nn} \neq 0$ for all n. If A is normal and $y_n = \Sigma_k a_{nk} x_k$ then it is easy to solve for x: $x_n = \Sigma_k b_{nk} y_k$, say. One finds also that $b_{nn} = 1/a_{nn}$.

Theorem 15. *Every normal co-null matrix is stronger than convergence, i.e. sums at least one divergent sequence.*

Proof. Let A be normal and co-null. The inverse matrix then exists and will be denoted by B. Write

$$b = \left(\sum_{k=0}^{n} b_{nk} \right)_{n \in N} \quad \text{and} \quad b_k = (b_{nk})_{n \in N} \quad (k = 0, 1, \ldots).$$

Then $A_n(b) = 1$ and $A_n(b_k) = \delta_{nk}$, so that the sequences b and b_k are in (A). Putting $\|x\| = \sup_n |A_n(x)|$, where $x \in (A)$, we see that (A) is a normed linear space. We can characterize the dual space $(A)^*$ by methods similar to those employed for c^* (see chapter 4). It is found that $f \in (A)^*$ if and only if there is a number t and a sequence $(t_k) \in l_1$ such that

$$f(x) = t \lim_n A_n(x) + \Sigma_k t_k A_k(x) \tag{20}$$

for all $x \in (A)$. When $f \in (A)^*$ we have in fact $t = t(f) = f(b) - \Sigma f(b_k)$ and $t_k = t_k(f) = f(b_k)$, where b, b_k are defined as above.

Suppose now that A is not stronger than convergence. Then $(A) = c$, and we shall show that $f(x) = \lim x_n$ defines an element f of $(A)^*$. Obviously f is linear on (A). Also, for $x \in c$,

$$|x_n| = |\Sigma_k x_k \delta_{nk}| = \left| \Sigma_k x_k \sum_{r=k}^{\infty} b_{nr} a_{rk} \right|$$

$$\leqslant \sum_{r=0}^{\infty} |b_{nr}| \left| \sum_{k=0}^{r} a_{rk} x_k \right|$$

$$\leqslant \|B\| \, \|x\|. \tag{21}$$

The manipulations are justified since $B = A^{-1}$ is conservative, as we now show. If $x \in c$ then $A(A^{-1}x) = (AA^{-1})x = x$ and so $A^{-1}x \in (A) = c$, whence A^{-1} is conservative. From (21) we get

$$|f(x)| = |\lim x_n| \leqslant \|B\| \, \|x\|,$$

so that f is bounded on (A). Now by (20) we have

$$f(e) - t \lim_n A_n(e) = \sum_{r=0}^{\infty} t_r \sum_{k=0}^{r} a_{rk}$$

$$= \sum_{k=0}^{\infty} \sum_{r=k}^{\infty} t_r A_r(e_k)$$

$$= \sum_{k=0}^{\infty} [f(e_k) - t A(e_k)],$$

i.e. $1 - t \lim_n \Sigma_k a_{nk} = - t \Sigma_k \lim_n a_{nk}$, i.e. $1 = t\chi(A)$, which is impossible since $\chi(A) = 0$. Thus the supposition that $(A) = c$ is false and so $c \subset (A)$ strictly, which proves the theorem.

There is a simple necessary condition for a γ-matrix to be stronger than convergence, which we now give.

Theorem 16. *If $A \in (\gamma, c; P)$ sums at least one divergent series then* $\lim_k a_{nk} = 0$ *for* $n = 0, 1, \ldots$.

Proof. Let Σb_k be divergent and summable A. Then $\Sigma a_{nk} b_k$ converges for each n. Hence $\Sigma a_{nk} b_k d_k$ converges for each n, whenever the sequence (d_k) is of bounded variation. Now $\sup_n \Sigma |\Delta a_{nk}| < \infty$ implies there exists $\lim_k a_{nk} = l_n$, $n = 0, 1, \ldots$. Suppose there exists n such that $l_n \neq 0$. Then for this n we have $|a_{nk}| \geqslant |l_n|/2$ for $k > M$ and so $(d_k)_{k>M} = (1/a_{nk})_{k>M}$ is of bounded variation. Hence $\sum_{k>M} a_{nk} b_k d_k$ converges, i.e. Σb_k converges, contrary to the fact that Σb_k diverges. Thus $l_n = 0$ for every n.

We now very briefly consider absolute and strong summability. A matrix A in (l_1, l_1) will be called *absolutely conservative*. If A is in $(l_1, l_1; P)$ it will be called *absolutely regular*. Using theorem 5 we see that A is absolutely regular if and only if

$$\sup_k \Sigma_n |a_{nk}| < \infty \quad \text{and} \quad \Sigma_n a_{nk} = 1 \quad (k = 1, 2, \ldots).$$

For any matrix A we denote by $|A|$ the set of all sequences x such that $A_n(x) = \Sigma_k a_{nk} x_k$ exists for each n and $\Sigma_n |A_n(x)| < \infty$. We then say that $x \in |A|$ is absolutely summable by A (or summable $|A|$). One has $l_1 \subset |A|$ if and only if A is absolutely conservative.

Perhaps the simplest method of absolute summability is that given by the Cesàro mean of order 1. If

$$t_n = \frac{1}{n+1} \sum_{k=0}^{n} (n+1-k) a_k$$

and $t \in c$ then we say that Σa_k is summable $(C, 1)$. If $t \in BV$ then we say that Σa_k is summable $|C, 1|$. It is trivial that $|C, 1| \subset (C, 1)$, and since

$$t_n - t_{n-1} = \frac{1}{n(n+1)} \sum_{k=1}^{n} k a_k \quad (n \geqslant 1)$$

it is clear that $l_1 \subset |C, 1|$. More generally it is true that $l_1 \subset |C, k|$, $k > 0$ (see exercises 3).

Analogous to the situation for ordinary summability, absolutely regular methods of summability may be equivalent to absolute convergence. For example, it is not hard to show that $|R^*, \lambda, 1| = l_1$ if and only if $\liminf \lambda_{n+1}/\lambda_n > 1$. Here we say that Σx_k is summable $|R^*, \lambda, 1|$ if and only if $t \in BV$, where

$$t_n = \sum_{k=0}^{n} (1 - \lambda_k/\lambda_{n+1}) x_k.$$

$|R^*, \lambda, 1|$ is 'discontinuous' absolute Riesz summability of order 1. We say that Σx_k is summable $|R, \lambda, k|$, $k > 0$ (where $|R, \lambda, k|$ is the 'continuous' case) if $C^k(U)$ is a function of bounded variation on $[\lambda_0, \infty)$, where

$$C^k(U) = \sum_{\lambda_n < U} (1 - \lambda_n/U)^k x_n.$$

By analogy with Steinhaus' theorem (theorem 14) one might be inclined to think that no absolutely regular matrix could sum all conditionally convergent series (the analogy is absolutely regular analogous to Toeplitz and conditionally convergent analogous to bounded, presumably). However, it turns out that there are absolutely regular matrices which do sum all conditionally convergent series. For example, $a_{1k} = 1$, $a_{nk} = 0$ $(n > 1, k = 1, 2, ...)$. See the exercises for some further results.

Finally, we turn to strong summability. For any matrix A and any sequence $p = (p_k)$ such that $p_k > 0$ we say that $x = (x_k)$ is summable $[A, p]$ (or strongly summable (A, p)) to l, written $x_k \to l[A, p]$, if $A_n(|x - le|^p)$ exists for each n and tends to zero as $n \to \infty$. In connection with strong summability we shall write

$$A_n(|x|^p) = \Sigma_k a_{nk} |x_k|^{p_k}.$$

If $A_n(|x|^p) = O(1)$ we say that x is strongly bounded by (A, p). Also, $[A, p]$ denotes the set of all x which are summable $[A, p]$, $[A, p]_\infty$ the set of x strongly bounded by (A, p) and $[A, p]_0$ the set of x such that $A_n(|x|^p) = o(1)$. Some well-known spaces are obtained by specializing A. For example, if $a_{nk} = 1$ $(1 \leqslant k \leqslant n)$, $a_{nk} = 0$ $(k > n)$, then $[A, p]_\infty = l(p)$. By $l_\infty(p)$ we denote $[I, p]_\infty$, where I is the unit matrix. If $A = (C, 1)$ then $x_k \to l[C, 1, p]$ means that

$$\frac{1}{n} \sum_{k=1}^{n} |x_k - l|^{p_k} \to 0.$$

We now prove three results concerning strong summability.

Theorem 17. *Let $p = (p_k)$ be bounded and A non-negative. Then $[A, p]_0$, $[A, p]$, $[A, p]_\infty$ are linear spaces.*

Proof. Just consider $[A, p]_0$; the others are similar. If $H = \sup p_k$ then

$$|a_k + b_k|^{p_k} \leqslant C(|a_k|^{p_k} + |b_k|^{p_k}),$$

where $C = \max(1, 2^{H-1})$. Also, $|\lambda|^{p_k} \leqslant \max(1, |\lambda|^H)$, and we write $\Lambda = \sup |\lambda|^{p_k}$, $M = \sup |\mu|^{p_k}$. Hence

$$\limsup A_n(|\lambda x + \mu y|^p) \leqslant C\Lambda \limsup A_n(|x|^p) + CM \limsup A_n(|y|^p).$$

The result follows.

For some matrices A the boundedness of p is necessary as well as sufficient for $[A, p]_0$, etc., to be linear spaces. As an example consider the space $l(p)$. If $l(p)$ is a linear space then $2x \in l(p)$ whenever $x \in l(p)$, i.e. $\Sigma 2^{p_k} |x_k|^{p_k} < \infty$ whenever $\Sigma |x_k|^{p_k} < \infty$. It follows readily that $p_k = O(1)$.

Theorem 18. *For any non-negative matrix A and any bounded sequence p, $[A, p]_0$ is a topological linear space, paranormed by*

$$g(x) = \sup_n (\Sigma_k a_{nk} |x_k|^{p_k})^{1/M},$$

where $M = \max(1, \sup p_k)$.

Proof. Clearly $g(\theta) = 0$, $g(x) = g(-x)$. That g is subadditive on $[A, p]_0$ follows by an argument similar to that for $l(p)$ given in chapter 2, section 1. We must check continuity of multiplication. For any complex λ we have $|\lambda|^{p_k} < \max(1, |\lambda|^H)$, where $H = \sup p_k < \infty$, so $g(\lambda x) < (\sup |\lambda|^{p_k})^{1/M} g(x)$ on $[A, p]_0$, whence $\lambda \to 0$, $x \to \theta$ imply $\lambda x \to \theta$, and also $x \to \theta$, λ fixed imply $\lambda x \to \theta$. Now let $\lambda \to 0$, x fixed. For $|\lambda| < 1$ we have $A_n(|\lambda x|^p) < \epsilon$ for $n > N(\epsilon)$. Also, for $1 \leqslant n \leqslant N$, since $\Sigma_k a_{nk}|x_k|^{p_k} < \infty$, there exists M such that

$$\sum_{k=M}^{\infty} a_{nk} |\lambda x_k|^{p_k} < \epsilon.$$

Taking λ small enough we then have $A_n(|\lambda x|^p) < 2\epsilon$ for all n, whence $g(\lambda x) \to 0$ as $\lambda \to 0$. This proves the theorem.

In our last theorem we consider the question of strong regularity. We say that the pair (A, p) is *strongly regular* if $x_k \to l$ implies $x_k \to l[A, p]$. The problem of finding exact conditions on A and p for strong regularity would seem to be quite difficult. By restricting p to begin with we do however have the following result.

Theorem 19. *Let m, M be constants such that $0 < m \leqslant p_k \leqslant M$. Then (A, p) is strongly regular if and only if $A \in (c_0, c_0)$.*

Proof. For the sufficiency, since $p_k \geqslant m > 0$, we have $x_k \to l$ implies $|x_k - l|^{p_k} \to 0$. Hence, $A \in (c_0, c_0)$ implies $x_k \to l[A, p]$.

Now suppose $x_k \to l$ implies $A_n(|x - le|^p) \to 0$. Then $|x_k - l|^{q_k} \to 0$ implies $\Sigma a_{nk}|x_k - l| \to 0$, where $q_k = 1/p_k$. Since $q_k \geqslant 1/M > 0$, $x_k \to l$ implies $|x_k - l|^{q_k} \to 0$. Hence, by decomposing $x_k - l$ into its real and imaginary parts and then these into their positive and negative parts we see that $x_k \to l$ implies $\Sigma a_{nk}(x_k - l) \to 0$, whence $A \in (c_0, c_0)$.

We remark that $p_k \leqslant M$ is superfluous in the sufficiency and $p_k \geqslant m$ is superfluous in the necessity.

Exercises 3

1. Prove that the Abel method is not weaker than any Cesàro method, i.e. show that $\Sigma a_n = s(C, r)$, $r > -1$, implies $\Sigma a_n = s$ (Abel). As a hint, note that

$$\Sigma a_n x^n = (1-x)^{r+1} \Sigma A_n^r x^n t_n^r,$$

where t_n^r is the Cesàro mean of order r. In fact the Abel method is stronger than all Cesàro methods (consider Σa_n where

$$\exp\{(1+x)^{-1}\} = \Sigma a_n x^n).$$

2. Prove that $a_{nk} = 2(n-k+1)n^{-1}(n+1)^{-1}$ $(1 \leqslant k \leqslant n)$, $a_{nk} = 0$ $(k > n)$, is a Toeplitz matrix. Find whether the sequence $((-1)^{k+1}k)$ is summable A.

3. (i) Prove that, to each Toeplitz matrix A corresponds the γ-matrix B, given by

$$b_{nk} = \sum_{p=k}^{\infty} a_{np}.$$

   (ii) Let B be a γ-matrix. Prove that $a_{nk} = b_{nk} - b_{n,\,k+1}$ is Toeplitz if and only if $\lim_n \lim_k b_{nk} = 0$.

4. Prove the following theorem of Steinhaus type: Given any γ-matrix B, there is always a series with bounded partial sums which is not summable B.

5. Prove that $a_{nk} = (e^{-n}n^{k+1})/(k+1)$ $(n,k = 1,2,...)$ is Toeplitz and show that the corresponding γ-matrix B (see question 3 (i)) is given by

$$b_{nk} = \frac{1}{k!} \int_0^n e^{-t} t^k \, dt.$$

6. We say that Σa_k is summable $|C,r|$ or $a \in |C,r|$ if the Cesàro mean t_n^r has bounded variation, i.e. $\Sigma |t_n^r - t_{n-1}^r| < \infty$. Prove that $l_1 \subset |C,r|$ for all $r > 0$.

7. Prove that $\Sigma(-1)^k k^{-1}$ is summable $|C,1|$ but that $\Sigma(-1)^k/\log(k+1)$ is not.

8. Define the transformation $B_n(a) = \Sigma_k b_{nk} a_k$ by $B_1(a) = a_1$ and

$$B_n(a) = \frac{n(n-2)}{(n-1)^2} a_{n-1} + \frac{a_n}{n^2} \quad (n \geqslant 2).$$

   Show that B is absolutely regular and that $\Sigma_n |B_n(a)| = \infty$ whenever Σa_k is conditionally convergent. Prove also that B sums absolutely a series for which $a_n \neq O(1)$.

9. Prove that every column bounded absolutely regular matrix sums every conditionally convergent series to which it applies. Column bounded means $a_{nk} = 0$ for $n \geqslant r$, where r is independent of k. Also, A applies to Σx_k means that $\Sigma a_{nk} x_k$ converges for each n.

10. In connection with strong summability, show that $l(p)$ can be regarded both as an $[A,p]_\infty$ space and as a $[B,p]_0$ space, i.e. find matrices A,B such that $l(p) = [A,p]_\infty = [B,p]_0$. For bounded p, prove that $l(p)$ is a topological linear space.

11. Let $0 < p_k \leqslant 1$ and write $[I,p]_\infty = l_\infty(p)$, where I is the unit matrix. Prove that $l_\infty(p)$ is paranormed by $g(x) = \sup |x_k|^{p_k}$; $x \in l_\infty(p)$, if and only if $\inf p_k > 0$.

12. If A is non-negative and $0 < \inf p_k \leqslant \sup p_k < \infty$, prove that $[A,p]_\infty$ is a topological linear space under the paranorm of theorem 18, section 3.

13. Prove that $[I,p]$, I the unit matrix, is a linear space if and only if p is bounded.

14. Define $a_{nk} = 1/n$ $(1 \leqslant k \leqslant n)$, $a_{nk} = 1$ $(k > n)$ and take $p_k = k$, for all $k \geqslant 1$. Prove that (A,p) is strongly regular but $A \notin (c_0,c_0)$. Compare with theorem 19, section 3.

4. Tauberian theorems

In this brief section we are merely able to introduce the vast field of Tauberian theory. For further details of the deeper Tauberian theorems the reader is referred to Hardy's *Divergent Series* and to the periodical literature cited there. Our concern will be with a fairly recent development of Tauberian theory which deals with the so-called *Tauberian constants*.

First we trace lightly the historical development. Abel proved in 1826 that if $\Sigma a_n = s$ then $\Sigma a_n = s$ (Abel), i.e. every convergent series is Abel summable to the same sum. As we know the converse is false in general, e.g. $\Sigma(-1)^n = 1/2$ (Abel) but $\Sigma(-1)^n \neq 1/2$. When is the converse true? By restricting the a_n, it was first shown by Tauber in 1897, that if $\Sigma a_n = s$ (Abel) and $na_n \to 0$ $(n \to \infty)$ then $\Sigma a_n = s$. Thus the condition $na_n \to 0$, now called a Tauberian condition, allows us to deduce convergence from Abel summability. Tauber also proved a related theorem, called Tauber's second theorem. Both his theorems are displayed for reference:

$$T_1: \quad \Sigma a_n = s \text{ (Abel)} \quad \text{and} \quad na_n \to 0 \quad \text{imply } \Sigma a_n = s.$$

$$T_2: \quad \Sigma a_n = s \text{ (Abel)} \quad \text{and} \quad na_n \to 0(C, 1) \quad \text{imply } \Sigma a_n = s.$$

Of course T_1 is a consequence of T_2, since $na_n \to 0$ implies $na_n \to 0(C, 1)$. In 1911, Littlewood initiated the deeper Tauberian theory by proving T_1 with $na_n \to 0$ (i.e. $na_n = o(1)$) replaced by $na_n = O(1)$, i.e. (na_n) bounded. Generally speaking a Tauberian theorem is a result which allows us to deduce convergence from summability, usually by putting an order condition on the terms of the series being summed. A very general Tauberian theory was created by Wiener in the 1930s and he used this theory to give a proof of the prime number theorem (Wiener, 1932).

Between the o-results and the O-results of Tauberian theory there lies an intermediate class of results which sharpen the o-results and are 'best possible' in a sense to be defined. These intermediate results are not as deep as the O-results but they are a worthwhile addition to the subject of Tauberian theory. The first of these intermediate results is due to Hadwiger (1944):

Theorem 20. *There exists an absolute constant M_1 such that*

$$\limsup_n \left| \sum_{k=1}^{\infty} a_k x_n^k - \sum_{k=1}^{n} a_k \right| \leqslant M_1 \limsup_n n |a_n|, \tag{22}$$

where $x_n = 1 - 1/n$.

The beauty of this result would seem to lie in the fact that there is no condition on (a_n). It is clear that Tauber's first theorem is a consequence of Hadwiger's theorem. A generalization of Tauber's second theorem is due to Wintner (1947):

Theorem 21. *There exists an absolute constant M_2 such that the left-hand side of (22) is less than or equal to*

$$M_2 \limsup_n n^{-1} \left| \sum_{k=1}^{n} k a_k \right|. \tag{23}$$

Wintner's result thus implies Tauber's second theorem. It was then shown by Hartman (1947) that the best values for the constants M_1 and M_2 were $M_1 = \gamma + 2\beta$, $M_2 = M_1 + 2/e$, where γ is Euler's constant $0.57721\ldots$ and

$$\beta = \int_1^\infty e^{-u} u^{-1} \, du = 0.21938\ldots.$$

By the best value we mean, for example,

$$M_1 = \max \left\{ (\limsup t_n)/(\limsup n \, |a_n|) \right\},$$

where the maximum is taken over all a such that $\limsup n \, |a_n| > 0$, and t_n denotes the expression in (22). The best values M_1 and M_2 are called *Tauberian constants*.

In this section we confine ourselves to proving a single general theorem on Tauberian constants. The best values of Hartman can then be obtained as a special case and with a minimum of calculation. We shall give, in theorem 22, necessary and sufficient conditions for the existence of Tauberian constants for general matrix transformations, with the Tauberian condition

$$A = \limsup_n n^{-1} \left| \sum_1^n k a_k \right| < \infty. \tag{24}$$

The condition $\limsup_n n \, |a_n| < \infty$ gives rise to a similar theorem (see exercises 5). Before the theorem we note a lemma. It may be proved by methods similar to those used in proving theorem 6, section 1, chapter 7. The details are left as an exercise.

Lemma. *Let $a_{nk} \to 0$ $(n \to \infty, k \text{ fixed})$ and*

$$\limsup_k |a_{nk}| = C < \infty.$$

Then $\qquad \limsup_n |\Sigma_k a_{nk} s_k| \leqslant C \limsup_n |s_n|,$

whenever $\lim \sup |s_n| < \infty$. *Also, there exists a sequence* (s_k) *with* $\lim \sup |s_n| = 1$ *such that*

$$\lim \sup_n |\Sigma_k a_{nk} s_k| = C.$$

In the theorem below we write $\Delta a_{nk} = a_{nk} - a_{n,\,k+1}$ for any matrix (a_{nk}). All sums run from $k = 1$ to $k = \infty$, unless otherwise indicated and upper limits are taken as $n \to \infty$.

Theorem 22. *Let A be given by* (24). *Then*

$$B_n(a) = \Sigma b_{nk} a_k, \quad C_n(a) = \Sigma c_{nk} a_k$$

exist for each n and there is a constant M such that

$$\lim \sup |B_n(a) - C_n(a)| \leqslant MA, \tag{25}$$

whenever $A < \infty$, *if and only if*

(a) $|b_{nk}| + |c_{nk}| \to 0$ ($k \to \infty$, n *fixed*),

(b) $b_{nk} - c_{nk} \to 0$ ($n \to \infty$, k *fixed*),

(c) $\Sigma k(|\Delta(b_{nk}/k)| + |\Delta(c_{nk}/k)|) < \infty$ (*each n*),

(d) $D = \lim \sup \Sigma k |\Delta(\{b_{nk} - c_{nk}\}/k)| < \infty$.

When (a)–(d) *hold we may take* $M = D$, *and this constant D is then best possible.*

Proof. *Sufficiency.* Let us write $n t_n = \overset{n}{\underset{k=1}{\Sigma}} k a_k$, so that for each n, whenever (a) and (c) hold and $A < \infty$,

$$\overset{p}{\underset{k=1}{\Sigma}} b_{nk} a_k = b_{np} t_p + \overset{p-1}{\underset{k=1}{\Sigma}} k \Delta(b_{nk}/k) t_k$$

$$\to \Sigma_k \Delta(b_{nk}/k) t_k \quad (p \to \infty). \tag{26}$$

The same is true with C in place of B. Hence

$$B_n(a) - C_n(a) = \Sigma k \Delta(\{b_{nk} - c_{nk}\}/k) t_k.$$

Thus, when (b), (d) hold we may apply the lemma and obtain (25) with $M = D$.

Necessity. If $B_n(a)$ exists whenever $A < \infty$, then (fixing n until further notice) by Toeplitz' theorem applied to the transformation of (t_k) given by (26), $\Sigma k |\Delta(b_{nk}/k)| < \infty$. \tag{27}

Since $\lim \sup_n |a_n| < \infty$ implies $A < \infty$, the existence of $B_n(a)$ implies that

$$\Sigma |b_{nk}|/k < \infty. \tag{28}$$

Now (27) implies that $b_{nk}/k \to l_n$ $(k \to \infty)$, whence

$$\sum_{k=1}^{p} b_{nk}/k \sim p l_n \quad (p \to \infty)$$

which is contrary to (28), unless $l_n = 0$. Hence

$$|b_{np}| = \left| p \sum_{k=p}^{\infty} \Delta(b_{nk}/k) \right| \leqslant \sum_{k=p}^{\infty} k |\Delta(b_{nk}/k)| \to 0 \quad (p \to \infty).$$

The same holds for C, whence (a) and (c) are necessary. The necessity of (b) follows from (25) on taking $a_k = 1$, $a_n = 0$ $(n \neq k)$, $k = 1, 2, \ldots$. We now have

$$B_n(a) - C_n(a) = \Sigma k \Delta(\{b_{nk} - c_{nk}\}/k) t_k,$$

for each n, whenever $A < \infty$. Now it is clear that the space X of all sequences $a = (a_k)$ such that $A < \infty$ is a Banach space with norm $\|a\| = \sup_n |t_n|$. Also, each f_n defined by $f_n(a) = B_n(a) - C_n(a)$ is in X^* and one readily finds that

$$\|f_n\| = \Sigma k |\Delta(\{b_{nk} - c_{nk}\}/k)|.$$

By hypothesis we have $\lim \sup |f_n(a)| \leqslant MA < \infty$ on X, whence by the Banach–Steinhaus theorem

$$D = \lim \sup \|f_n\| < \infty,$$

which is (d). This proves the theorem.

As a corollary to theorem 22 we shall obtain Hartman's Tauberian constant $\gamma + 2\beta + 2/e$, which is the best value of M_2 in theorem 21. Take $b_{nk} = (1 - 1/n)^k = x^k$, where $x = x_n = 1 - 1/n$, and $c_{nk} = 1$ $(1 \leqslant k \leqslant n)$, $c_{nk} = 0$ $(k > n)$. Then it is straightforward to check that (a)–(d) of theorem 22 hold and that

$$D = \lim \sup_n \left\{ 2x^{n+1} - 1/n - \log n - \sum_{1}^{n} 1/k + 2 \sum_{n+1}^{\infty} x^{k+1}/(k+1) \right\}. \tag{29}$$

Now it is not hard to show that

$$\sum_{n+1}^{\infty} x^{k+1}/(k+1) \to \beta = \int_{1}^{\infty} e^{-u} u^{-1} du \quad (n \to \infty)$$

(see exercise 4). Hence $D = 2/e + \gamma + 2\beta$, since

$$1 + 1/2 + \ldots + 1/n = \log n + \gamma + o(1),$$

as is well known from elementary analysis. Thus we have obtained the Tauberian constant D for the Abel method. It is of interest to note that in this case D exists as a limit not just as an upper limit.

A result similar to that of theorem 22 involves the Tauberian condition $\limsup n\,|a_n| < \infty$. The proof follows the same lines but is rather simpler (see exercise 5).

Exercises 4

1. Prove Tauber's theorems T_1 and T_2, without assuming the theorems of Hadwiger and Wintner.
2. Prove the lemma which precedes theorem 22, section 4.
3. Verify equation (29) of section 4.
4. Let $r = r(n)$ be such that $0 < r < 1$ and $-n \log r \to 1$ as $n \to \infty$. Prove that

$$\sum_{k=n+1}^{\infty} r^k k^{-1} = \int_{-n \log r}^{\infty} n^{-1} (e^{t/n} - 1)^{-1} e^{-t} dt$$

and that this last integral tends to

$$\beta = \int_{1}^{\infty} e^{-t} t^{-1} dt$$

as $n \to \infty$.

5. Prove the companion to theorem 22: $B_n(a)$, $C_n(a)$ exist for each n and there is a constant M such that

$$\limsup_n |B_n(a) - C_n(a)| \leqslant MA',$$

whenever $\qquad A' = \limsup_n n\,|a_n| < \infty,$

if and only if

$\qquad (b)' \quad b_{nk} - c_{nk} \to 0 \quad (n \to \infty,\ k \text{ fixed}),$

$\qquad (c)' \quad \Sigma k^{-1}(|b_{nk}| + |c_{nk}|) < \infty \quad (\text{each } n),$

$\qquad (d)' \quad D = \limsup_n \Sigma k^{-1} |b_{nk} - c_{nk}| < \infty.$

When the conditions hold we may take $M = D$, which is then best possible.

5. Some problems for further study

Here we present a collection of questions for the reader who may be interested in continuing the study of the topics of this and other chapters. Some of the questions are of a fairly elementary standard—these are unstarred. Others have a known solution, which has sometimes taken years to find. Where problems are considered fairly difficult I have starred them, e.g. 2*. Certain problems begin thus: Find the necessary and sufficient conditions for.... These problems will often, as far as I know, bring the reader up to the present state of knowledge. A few questions bear a double star—these should prove quite stimulating.

The purpose of this section will be best served if the reader is led to formulate (and solve!) new problems for himself.

1.** A normed space with a basis is separable. Some separable Banach spaces have a basis. Does every separable Banach space have a basis? (The answer is not known.)

2.* A *rotation* U (about θ) in a linear metric space X is a surjective isometry of X into itself such that $U(\theta) = \theta$.

Prove that if X is a real normed space then every rotation is necessarily linear (see Banach, 1955, p. 166).

3.* Prove that the general rotation in l_p ($1 \leqslant p \neq 2$) takes the form $y_n = \epsilon_n x_{\phi(n)}$, where $|\epsilon_n| = 1$, ϕ is a permutation of N and $y = U(x) = (y_n)$. (See Banach, 1955, p. 178.)

4.** Find the form of the general rotation in $l(p)$. (The answer is not known.)

5. If X is a Banach space then $X \times X = X^2$ is a Banach space with $\|z\| = \|x\| + \|y\|$, where $z = (x, y) \in X^2$. We define

$$(x, y) + (x', y') = (x + x', y + y') \quad \text{and} \quad \lambda(x, y) = (\lambda x, \lambda y).$$

Prove that c is isomorphic to c^2 and l_p is isomorphic to l_p^2 ($p \geqslant 1$).

6.* Prove that $C[0, 1]$ is isomorphic to $C[0, 1] \times c$.

7. Let X be a sequence space and E a subset of X. Prove that

$$E^\dagger \equiv \{x \mid \Sigma |x_k y_k| < \infty, \text{ for all } y \in E\}$$

is a subspace of X. $E^\dagger$ is called the *Köthe–Toeplitz dual* of E. Note that we are using the same notation as for the algebraic dual of a linear space—this should do no harm.

Prove that
$$c^\dagger = l_1, \quad l_1^\dagger = l_\infty, \quad l_p^\dagger = l_q \quad (1 < p < \infty).$$
What is $l_\infty^\dagger$?

8.** Find $l(p)^\dagger$ (see question 7 for the definition).

9. Prove that $A \in (\gamma, \gamma; P)$ if and only if

$$\sup_n \sum_k \left| \sum_{r=1}^n \Delta a_{rk} \right| < \infty, \quad \sum_n a_{nk} = 1 \quad (k = 1, 2, \ldots).$$

10. Show that $A \in (BV, l_\infty)$ if and only if

$$M = \sup_{n, k} |a_{n1} + a_{n2} + \ldots + a_{nk}| < \infty.$$

Prove also that $A \in (BV, c)$ if and only if $M < \infty$ and there exists

$$\lim_n \sum_{k=p}^{\infty} a_{nk} \quad (p = 1, 2, \ldots).$$

11.* Find the necessary and sufficient conditions for $A \in (l_1, c)$. (See Cooke, 1950.)

12.** Find the necessary and sufficient conditions on A for $A \in (l_p, l_s)$, $p > 0$, $s > 0$ (include l_∞ as well). Even the case $p = s = 2$ has not been solved (see theorem 9, section 1, chapter 7).

13. Let $(C, -1)$ be the set of all sequences x such that $x \in \gamma$ and $nx_n \to 0$. A series Σx_k is said to be summable $(C, -1)$ if $x \in (C, -1)$. Write $t_n = s_n + nx_n$ with $s_n = x_0 + x_1 + \ldots + x_n$. Show that $x \in (C, -1)$ if and only if $t \in c$.

Prove also that a matrix $A \in ((C, -1), c; P)$ if and only if A is representable as

$$a_{nk} = b_{nk} + k \Delta b_{nk},$$

where B is a γ-matrix. Show the representation is unique. (See Maddox, 1965, 1966a.)

14. Find whether (l_1, l_p), $1 \leqslant p \leqslant \infty$, is a Banach space with

$$\|A\| = \sup_k (\Sigma_n |a_{nk}|^p)^{1/p}.$$

Is (l_1, l_p) an algebra under the various products defined in section 2, chapter 7?

15.* Can you define a convolution type operation in (l_1, l_1) such that (l_1, l_1) becomes a Banach algebra with identity?

16. For matrices A, B, write $A \supset B$ if the B-summability of a sequence implies its A-summability to the same limit. Prove that $A \supset (R, \lambda, 1)$ if and only if A is Toeplitz and

$$\sup_n \Sigma_k \lambda_{k+1} |\Delta(a_{nk}/\Delta\lambda_k)| < \infty.$$

(See Maddox, 1964.) In this question, $(R, \lambda, 1)$ denotes the Riesz mean of type λ and order 1 (section 1, chapter 7).

17.** Find necessary and sufficient conditions on a matrix A for $A \supset (R, \lambda, k)$. (See 16 above and Maddox, 1964, for a partial solution.) (The answer is not known.)

18.** Find necessary and sufficient conditions for $A \supset (C, k)$. (See Cooke, 1952.)

19.* (Fejér's theorem). If f is periodic 2π and integrable on $[0, 2\pi]$ and such that $f(x_0 + 2t) + f(x_0 - 2t) \to 2s(x_0)$ as $t \to 0$, prove that the Fourier series $a_0/2 + \sum_{n \geqslant 1} (a_n \cos nx_0 + b_n \sin nx_0)$ of f is summable $(C, 1)$ to $s(x_0)$.

If, in addition, f is continuous on $[0, 2\pi]$ prove that the Fourier series is uniformly summable $(C, 1)$ to f on $[0, 2\pi]$ and so everywhere. Deduce the Weierstrass approximation theorem: the set of polynomials is dense in $C[0, 2\pi]$ (or $C[0, 1]$). (See Knopp's book.)

20. If Σa_n is summable (C, k), $k > -1$, prove that $a_n = o(n^k)$. This is the 'limitation theorem' for Cesàro means.

21.* If $\Sigma a_n = s$ and $na_n = O(1)$, prove that $\Sigma a_n = s(C, -1+d)$ for $d > 0$ (see Hardy, 1945, theorem 45).

22.** Let $k > 0$ be fixed. Prove that the Riesz mean (R, λ, k) is equivalent to convergence if and only if

$$\Lambda_n = \lambda_{n+1}/(\lambda_{n+1} - \lambda_n) = O(1)$$

(see Hardy and Riesz, 1915; and Kuttner, 1965).

23.** Let $k > 0$ be fixed. Find the necessary and sufficient condition on λ for the absolute Riesz mean $|R, \lambda, k|$ to be equivalent to absolute convergence. The answer is not known, but it is conjectured that the condition is $\Lambda_n = O(1)$ (see 22, above). (For certain cases of this problem see Maddox, 1966b.)

24.* A is a lower-semi Toeplitz matrix such that

$$\liminf_n \left(|a_{nn}| - \sum_{k=1}^{n-1} |a_{nk}| \right) > 0.$$

Prove that A is equivalent to convergence (Agnew, 1952).

25. For $x \in (A)$, put $\|x\| = \sup_n |\Sigma_k a_{nk} x_k|$, where A is conservative, and identify members of (A) whose distance apart is zero. Prove that (A) is separable.

26.* Prove that if a conservative matrix sums a bounded divergent sequence then it sums an unbounded one (Wilansky's book, p. 231, question 31).

27. Let f, g be continuous on $[0, \infty)$ and $f(x) \sim Ax^a$, $g(x) \sim Bx^b$ $(x \to \infty)$, where A, B, a, b are constants and $a > -1$, $b > -1$. Prove that, as $x \to \infty$,

$$\int_0^x f(t) g(x-t) \, dt \sim AB \frac{\Gamma(a+1)\,(\Gamma(b+1)}{\Gamma(a+b+2)} x^{a+b+1}.$$

28. For Riesz means, prove that

$$(R, \lambda, k) \subset (R, \lambda, k') \quad \text{and} \quad |R, \lambda, k| \subset |R, \lambda, k'|$$

when $k' \geqslant k \geqslant 0$. (Use 27 above and the relation

$$C^{k+l}(x) = \frac{\Gamma(k+l+1)}{\Gamma(k+1)\Gamma(l)} \int_0^x (x-t)^{l-1} C^k(t) \, dt,$$

where

$$C^k(x) = \sum_{\lambda_n < x} (x - \lambda_n)^k a_n$$

for a series Σa_n.)

29.* Prove that $A \in (\gamma, l_1)$ if and only if $\Sigma_n |a_{n1}| < \infty$ and

$$\sup_k \sum_k | \sum_{n \in \sigma} \Delta a_{nk}| < \infty. \tag{1}$$

Here σ denotes a finite set of positive integers, F the class of all sets σ and in the above the supremum is taken over all elements of F (see Maddox, 1967a).

30. Show that $A \in (l_1, l_1; P) \cap (\gamma, l_1)$ if and only if A is absolutely regular and (1) of 29 holds.

31.** Find the necessary and sufficient conditions on A such that

(i) $A \in (BV, l_1)$,   (ii) $A \in (c, l_1)$,   (iii) $A \in (l_\infty, l_1)$.

32.** Try to find a result for absolutely conservative matrices analogous to theorem 15, section 3, chapter 7—replacing convergence by absolute convergence and divergent sequence by conditionally convergent series.

33. Let A be a non-negative matrix and p any positive sequence. Prove that $[A, p]_\infty$, $[A, p]$ and $[A, p]_0$ (see section 3, chapter 7) are absolutely convex subsets of the space of all sequences.

34. Let $p_k > 0$, $q_k > 0$ and $c_0(p)$ denote the set of all $x = (x_k)$ such that $|x_k|^{p_k} \to 0$. Prove that $c_0(q) \subset c_0(p)$ if and only if

$$\lim \inf p_k/q_k > 0.$$

35.** Find necessary and sufficient conditions on $p = (p_k)$ such that $x_k \to l$ implies $x_k \to l[C, 1, p]$.

36.   Let $0 < p < q$. Prove that $x_k \to l[A, q]$ implies $x_k \to l[A, p]$ when A is non-negative and in (l_∞, l_∞).

37. Let A be Toeplitz and suppose $0 < p_k \leqslant q_k$, with $q_k/p_k \to \infty$. Prove that $x_k \to l[A, q]$ does not generally imply $x_k \to l[A, p]$ (Maddox, 1967b).

38.* Prove that $l(p)^* = l_\infty(p)$ for $0 < p_k \leqslant 1$ (Simons, 1965).

39.** Characterize $[C, 1, p]^*$ for bounded p.

40.** Find the necessary and sufficient conditions on p such that the limit of a sequence summable $[C, 1, p]$ is unique.

41. (Kuttner's theorem.) Let $0 < p < 1$, $p_k = p$ for all k and let A be Toeplitz. Prove that there is always a sequence summable $[C, 1, (p_k)]$ which is not summable A. Compare with Steinhaus' theorem, section 3, chapter 7. For the proof of Kuttner's theorem use theorem 7, section 1, chapter 7 (see Maddox, 1968; Kuttner, 1946).

BIBLIOGRAPHY

AGNEW, R. P., Equivalence of methods for evaluation of sequences, *Proc. American Math. Soc.* **3** (1952), 550–6.

AHLFORS, L., *Complex Analysis* (McGraw-Hill, 1953).

BANACH, S., *Théorie des opérations linéaires* (Chelsea, New York, 1955).

COHEN, L. W. and ERLICH, G., *The Structure of the Real Number System* (Van Nostrand, 1963).

COOKE, R. G., *Infinite Matrices and Sequence Spaces* (Macmillan, 1950).

COOKE, R. G., On T-matrices at least as efficient as (C, r) summability, and Fourier-effective methods of summation, *J. London Math. Soc.* **27** (1952), 328–37.

GELFAND, I. M. and SHILOV, G. E., *Generalized Functions*, vol. I (Academic Press, 1964).

HADWIGER, H. Über ein Distanz-theorem bei der A-Limitierung, *Comm. Math. Helv.* **16** (1944), 209–14.

HALMOS, P. R., *Lectures on Ergodic Theory* (Chelsea, New York, 1956).

HARDY, G. H., *Divergent Series* (Oxford, 1949).

HARDY, G. H. and RIESZ, M., *The General Theory of Dirichlet's Series* (Cambridge Tract No. 18, 1915).

HARTMAN, P., Tauber's theorem and absolute constants, *Amer. J. of Math.* **69** (1947), 599–606.

HILBERT, D., *Grundzüge einer allgemeinen Theorie der linearen Integralgleichungen*, Leipzig, 1912.

JONES, D. S., *Generalized Functions* (McGraw-Hill, 1966).

KNOPP, K., *Theory and Application of Infinite Series* (Blackie, 1964).

KUTTNER, B., Note on strong summability, *J. London Math. Soc.* **21** (1946), 118–22.

KUTTNER, B., On discontinuous Riesz means of order 2, *J. London Math. Soc.* **40** (1965), 332–7.

LITTLEWOOD, J. E., The converse of Abel's theorem on power series, *P. London Math. Soc.* (2), **9** (1911), 434–83.

MADDOX, I. J., Some inclusion theorems, *Proc. Glasgow Math. Assn.* **6** (1964), 161–8.

MADDOX, I. J., Matrix transformations of $(C, -1)$, summable series, *Proc. Kon. Ned. Akad. van Wetensch.* A **68** (1965), 129–32.

MADDOX, I. J., Matrix transformations in a Banach space, *Proc. Kon. Ned. Akad. van Wetensch.* A **69** (1966), 25–9.

MADDOX, I. J., Note on Riesz means, *Quarterly J. of Math.* (2), **17** (1966), 263–8.

MADDOX, I. J., On theorems of Steinhaus type. *J. London Math. Soc.* **42** (1967), 239–44.

MADDOX, I. J., Spaces of strongly summable sequences, *Quarterly J. of Math.* (2), **18** (1967), 345–55.

MADDOX, I. J., On Kuttner's theorem, *J. London Math. Soc.* **43** (1968), 285–90.

ROBERTSON, A. P. and ROBERTSON, W. J., *Topological Vector Spaces* (Cambridge, 1964).

RUDIN, W., *Principles of Mathematical Analysis* (McGraw-Hill, 1964).

SCHWARTZ, L., *Théorie des distributions* (Paris, Hermann, 1950, 1951).

SIMONS, S., The sequence spaces $l(p_\nu)$ and $m(p_\nu)$, *P. London Math. Soc.* (3), **15** (1965), 422–36.

SUPPES, P. C., *Axiomatic Set Theory* (Van Nostrand, 1960).

TAYLOR, A. E., *Introduction to Functional Analysis* (Wiley, New York, 1958).

WIENER, N., Tauberian theorems, *Ann. of Math.* (2), **33** (1932), 1–100.

WILANSKY, A., *Functional Analysis* (Blaisdell, 1964).

WINTNER, A., On Tauber's theorem, *Comm. Math. Helv.* **20** (1947), 216–22.

ZYGMUND, A., *Trigonometrical Series* (Dover, 1955).

INDEX